Increasing the Durability of Paint and Varnish Coatings in Building Products and Construction

Woodhead Publishing Series in Civil and Structural Engineering

Increasing the Durability of Paint and Varnish Coatings in Building Products and Construction

Loganina Valentina Ivanovna

Doctor of Technical Sciences

Professor and Head of quality control and construction technologies at Penza State University of Architecture, Penza, Russia

ELSEVIER

WP
WOODHEAD PUBLISHING
An imprint of Elsevier

Woodhead Publishing is an imprint of Elsevier
The Officers' Mess Business Centre, Royston Road, Duxford, CB22 4QH, United Kingdom
50 Hampshire Street, 5th Floor, Cambridge, MA 02139, United States
The Boulevard, Langford Lane, Kidlington, OX5 1GB, United Kingdom

British Library Cataloguing-in-Publication Data
A catalogue record for this book is available from the British Library

Library of Congress Cataloging-in-Publication Data
A catalog record for this book is available from the Library of Congress

ISBN: 978-0-12-817046-5 (print)

ISBN: 978-0-12-817047-2 (online)

For information on all Woodhead Publishing publications
visit our website at https://www.elsevier.com/books-and-journals

Publisher: Matthew Deans
Acquisition Editor: Glyn Jones
Editorial Project Manager: John Leonard
Production Project Manager: Sojan P. Pazhayattil
Cover Designer: Greg Harris

Typeset by MPS Limited, Chennai, India

Contents

Regularities for formation of the quality of the external appearance of coatings

1

1.1 Evaluation of the quality of the appearance of paint and varnish coatings

Coatings for finishing facades with aesthetic protective functions must have a high-quality appearance. The quality of appearance refers to the presence of defects (inclusions, streaks, shagreen, strokes and scratches, waviness, variability) paintwork. The presence of defects on the surface of the paintwork determines the quality of appearance.

According to GOST 9.032-74 "Unified system of protection against corrosion and aging for coatings for paint and varnish, groups, technical requirements and designations," there are seven classes of the quality of the appearance of coatings on a metal substrate defined as follows:

I class—no defects can be allowed for high-gloss, glossy, semi-glossy and semi-matt. For matt coatings not more than 4 inclusions per m^2;

II–VII classes—are possible weed or individual inclusions, taking into account their number (pcs/m^2) depending on the length, width, diameter of the defect and the distance between them (mm), as well as risks and hatch.

In addition to the above defects, the quality classes III–YII include waviness, Y–YII streaks, and IY–YII offshade.

In Refs. [1,2], the quality of the appearance of protective-decorative coatings for cement concrete are evaluate IV–VII classes (Table 1.1).

As noted earlier, the surface quality of the paintwork is also determined by its roughness, that is, its surface profile. Methods involving fractal physics can be used to estimate the coating surface profile. The fractal dimension index D of the coating surface profile ranges between $1 < D < 2$ [3–6]. The greater the roughness of the coating, the more curved the coating profile and the greater the value of D. Thus, the fractal dimension D of the coating profile can serve as a criterion of its appearance quality, reflecting the presence of inclusions, streaks, and waviness.

In this chapter, we have attempted to evaluate the possibility of describing the coating surface quality using fractal dimension.

The paints were applied with a brush onto a mortar substrate in two layers with 24-hour intermediate drying. The coating profile was determined with a profile meter A1. The length of the coating surface profile was determined with a perambulator.

Increasing the Durability of Paint and Varnish Coatings in Building Products and Construction.
DOI: https://doi.org/10.1016/B978-0-12-817046-5.00001-2

Table 1.1 Permissible defects for coatings depending on their class

Coating class	Defect name	Norm for coatings		
		Satin gloss	Semimaturing	Matt
IV	(a) Inclusions: number of pcs/dm^2	1	1	1
	(b) Size, mm, not more	1.0	1.0	1.0
	(c) Distance between inclusions, mm, no more than	10	10	10
	(d) Shagreen	Allowed to		
	(e) Streaks	Not allowed		
	(f) Strokes, risks	Allowed individual		
	(g) Waviness, mm, not more than	2	2	2
	(h) Offshade	Not allowed		
Y	(a) Inclusions: number of pcs/dm^2	4	4	5
	(b) Size, mm, not more	2.0	2.0	2.0
	(c) Shagreen	Allowed to		
	(d) Streaks	Allowed Individual		
	(e) Strokes, risks	Allowed to		
	(f) Waviness, mm, not more than	2.5	2.5	2.5
	(g) Offshade	Not allowed		
YI	(a) Inclusions: number of pcs/dm^2	8	8	8
	(b) Size, mm, not more	3.0	3.0	3.0
	(c) Shagreen	Not allowed		
	(d) Streaks	Allowed individual		
	(e) Strokes, risks	Allowed to		
	(f) Waviness, mm, not more than	4.0	4.0	4.0
	(g) Offshade	Allowed to		

In addition, the coating surface quality was estimated with a gloss meter FB-2. The fractal dimension D of the coating surface profile was estimated using the geometrical method. For this purpose the image of the curve obtained by using a profile meter profilograph was covered by a grid consisting of squares with side L1. Then the number of the squares N curve N (L1) went through were counted. After changing the grid scale, we recounted the number of squares intersecting curve N (L2), N (L3) N (Ln). Then, in a log–log grid we plotted N (L), according to the inclination angle of which we determined the fractal dimension.

Table 1.2 shows the results of the coating surface quality estimation that was carried out in accordance with GOST 9.407-74 using the fractal dimension of the coating surface profile.

As can be seen, there is a correlation between the coating surface roughness and the appearance quality class. As the surface roughness increases the appearance quality class decreases and the fractal dimension D increases.

For example, the coating surface profile fractal dimension based on PF-115 with a coating surface roughness and appearance quality class of 0.2 m and Y is $D = 1.17$, while with the coating surface roughness and the appearance quality class, it is 0.75 mkm and YI is $D = 1.35$. Similar patterns were seen for other types of coatings.

Changes in the gloss and coating surface profile parameters on the sample were equal to 10 cm also prove the fact. When the coating surface profile fractal dimension increases, the gloss decreases and the numerical values of the profile perimeter increase.

Table 1.2 Coating appearance quality estimation

The paint	The coating surface roughness, R_a (mkm)	Fractal dimension, D	The surface profile perimeter (mm)	The coating surface quality kind	Gloss (%)
PF-115	0.22	1.17	106	Y	70.8
	0.75	1.35	116	YI	67.7
MA-15	0.2	1.075	102	Y	83.1
	0.48	1.125	109	YI	78
NC-123	0.2	1.1	105	Y	76.9
	0.75	1.36	121	YI	67.7
Acril, Universal	0.75	1.3	216	YI	66.2
	1.26	1.42	221	YII	64.2
Water-dispersible, facade	0.6	1.25	192	YI	66.2
	1.24	1.7	225	YII	62.2

The correlation between the length L of the coating surface profile on the sample length l and the fractal dimension D can be approximated as

$$L = lD^{bD} \tag{1.1}$$

where b is a constant factor.

In particular, for the coatings based on paint PF-115 the expression (1.1) is

$$L = 99.62D^{0.33D}$$

For paint MA-15-based coating

$$L = 99.08D^{0.638D}$$

For paint NC-based coating

$$L = 99.55D^{0.416D}$$

For acrylic paint-based coating

$$L = 115.28D^{1.47D}$$

For water-dispersible façade paint-based coating

$$L = 131D^{0.645D}$$

The results of the research suggest that at the fractal dimension D, equal $D = 1 - 1.09$, quality class is IY, at $D = 1.1 - 1.2$ the coating appearance quality class is Y, when $D = 1.21 - 1.4$—YI, and at $D = 1.41 - 1.99$—YII.

When applying the integral performance index to the coatings appearance fractal dimensions help estimate paint coatings quality more objectively.

1.2 Quality of the appearance of paint and varnish coatings

Research has shown that the quality of the appearance of coatings is determined by the nature of the bottling of the paint. In accordance with [7] bottling is considered as a rheological process, which can be described by the following expression:

$$h = \frac{b^2 f}{8\sigma} \tag{1.2}$$

where h is the stroke height

b is the width of stroke
f is the shift limit stress of the paint
σ is the surface tension of the paint

The following procedure was used to determine the paint's ability to flow on the surface of the substrate (applied by a brush). Five pairs of parallel strips of material were applied and the degree of spreading over the number of merged bands was determined. The strips were made with a special device. Estimation of the degree of spreading of the five pairs of parallel strips was determined using a 10-point scale.

To assess the spreading of paint a technique based on the determination of the surface profile of the coating was used. The following paints were used in the work: alkyd grade enamel PF-115, oil paint MA-15, enamel nitrocellulose paint of grade NC-123, acrylate paint class "universal," and acrylic water-dispersion (facade).

The paints were applied on the cement substrates in two layers using a brush with intermediate drying within 24 hours. The surface tension of the paint composition was determined by the drop method (stalagmometric method). From a special capillary—a stalagmometer—the same volumes of water and the test liquid or solution are squeezed out. The number of drops formed from the same volume of liquid is proportional to the density of this liquid and inversely proportional to its surface tension.

The value of the surface tension of paint was calculated from the formula

$$\sigma = K\frac{\rho}{n} \tag{1.3}$$

Distilled water with density $\rho_m^{20''-} = 0.9982\ \mathrm{g/cm^2}$ and surface tension $\sigma^{20} = 72.8\ \mathrm{mN/m}$ was used as the reference liquid.

The dynamic viscosity of the paint was determined by the Stokes method. To do this, the test liquid is poured into a graduated cylinder, in which a metal ball of known density ρ_{ball} and radius r is brought to the surface and lowered (without a push). As soon as the movement of the ball becomes uniform, the electric second-timer is turned on and the travel time t of the ball is determined. The viscosity of the liquid is given by the formula:

$$\eta = \frac{2}{9}gr^2\frac{(\rho_{\mathrm{ball}} - \rho_{\mathrm{paint}}) \times t}{l} \tag{1.4}$$

where g is the acceleration of gravity m/s^2

t is the time during which the ball passes the distance between the marks A and B, s
l is the distance between the marks A and B, cm
$\rho_{\mathrm{ball}}, \rho_{\mathrm{paint}}$ is the density of the ball and paint, g/cm^3, respectively

The substrates were made of glass and cement-sand mortar. The results of the studies are given in Table 1.3. Analysis of the data (Table 1.3) indicates that there is some slowing down of the time of restoration of the structure of paint when applied to the porous surface of the solution. For example, on a glass substrate, the time for restoring the structure of the PF-115 paint composition is 3 minutes, and on the cement substrate it is 5 minutes, and filling in both cases is satisfactory. Undoubtedly, the time to restore the structure of the paint composition depends both on the porosity of the substrate and on the rheological properties of the paint.

The results of evaluation of the surface profile of coatings indicate that with an increase in the surface tension of the paint composition, a lower quality of the appearance of the coating formed is observed. Thus, with the surface tension of an acrylate paint of the universal class equal to $s = 36.13$ mJ/m^2, the surface roughness of the coating is $R_a = 2.4$ mkm, and the fractal dimension is $D = 1.53$. Similar patterns are also seen for other types of coatings.

Continuing this research, the following experiment was performed to establish the relationship between the quality indices of the appearance of the coatings and the porosity of the cement substrate. The test paints were applied to the substrate in two layers with intermediate drying for 20 minutes. Before applying paint, the surface of the substrate was primed. During the testing samples with a volume porosity of 24%, 28%, and 32% were used. The results of the studies are given in Table 1.4.

According to ISO 1302, there are 12 classes of surface roughness (from N1 to N12). Analysis of the data (Table 1.4) shows that the high quality of the appearance of the coating (roughness class N2) can be obtained for the porosity of $P = 24\%$ only when using alkyd paint PF-115 with a dynamic viscosity of 0.001×10^3 Pa s. With other indicators of the viscosity of PF-115 paint with a given porosity, coatings with a roughness class N3 are formed. The use of cement substrates with porosity $P = 32\%$ does not ensure the production of coatings with a roughness class N2, regardless of the rheological properties of the paint compositions.

For aqueous paint the roughness index decreases with increasing surface tension to a value within the range of 48–55 mJ/m^2 (Fig. 1.1).

Further increase in surface tension (dilution of the paint with water) promotes an increase in roughness, that is, reduces the quality of the appearance of the coating.

Analysis of the data in Fig. 1.1 shows that the dependence of the surface roughness R_a on the surface tension can be approximated by an equation of the form

$$R_a = a + b\sigma_{st} + c\sigma_{st}^2 \tag{1.5}$$

where s_{nn} is the surface tension of the colorful composition.

Table 1.3 Effect of rheological properties of paints on the quality of the appearance of coatings

Name of paint	Surface roughness, R_a (mkm)	Fractal dimension of the surface of the coating, D	Surface tension of the paint composition (mJ/m^2)	Dynamic viscosity of the paint composition (Pa s)	Filling the colorful composition (min)
Alkyd paint PF-115	0.58	1.29	19.35	7.92	No more 5 / No more 5[a]
	0.4	1.18	18.37	6.86	No more 5 / No more 3[a]
	0.21	1.06	16.67	5.8	No more 5 / No more 3[a]
Oil paint MA-15	0.8	1.3	17.48	23.68	No more 5 / No more 3[a]
	0.69	1.1	16.93	14.8	No more 5 / No more 3[a]
	0.46	1.03	16.18	10.36	No more 5 / No more 3[a]
Nitrocellulose NC-123	0.78	1.3	27.09	14.02	No more 5 / No more 3[a]
	0.6	1.12	24.08	7.38	No more 5 / No more 3[a]
	0.32	1.06	22.12	6.39	No more 5 / No more 3[a]
Acryl water dispersed (façade) paint	3.01	1.4	37.37	40.04	More 15 / More 15[a]
	2.55	1.25	34.96	30.8	More 15 / More 15[a]
	1.85	1.1	31.88	21.56	More 15 / More 15[a]
Acrylate class Universal	2.4	1.53	36.13	33.44	More 15 / More 15[a]
	1.77	1.3	33.87	24.32	More 15 / More 15[a]
	1.44	1.1	30.96	15.2	More 15 / More 15[a]

[a]*Note*: The roughness of the coating was evaluated on a glass substrate.

Table 1.4 Quality of the appearance of the paint coatings as a function of the dynamic viscosity and porosity of the substrate

Porosity (%)	Type of paint								
	PF-115			MA-15			WD		
	Dynamic viscosity, η (Pa s)			Dynamic viscosity, η (Pa s)			Dynamic viscosity, η (Pa s)		
	$h_1 = 0.001 \times 10^3$	$h_2 = 0.00065 \times 10^3$	$h_3 = 0.00026 \times 10^3$	$h_1 = 0.0026 \times 10^3$	$h_2 = 0.0026 \times 10^3$	$h_3 = 0.0014 \times 10^3$	$h_1 = 0.0347 \times 10^3$	$h_2 = 0.02317 \times 10^3$	$h_3 = 0.013 \times 10^3$
24	0.98(N2)	2.05(N3)	2.7(N3)	3.53(N3)	4.14(N4)	4.54(N4)	2.87(N3)	3.13(N3)	3.38(N3)
28	1.7(N2)	2.26(N3)	3.0(N3)	5.61(N4)	6.5(N4)	7.56(N4)	3.1(N3)	3.37(N3)	3.55(N3)
32	2.38(N3)	2.86(N3)	3.5(N3)	7.76(N4)	8.99(N5)	9.26(N5)	3.32(N3)	3.55(N3)	4.2(N4)

Note: The values of the roughness class of the surface are given in parentheses.

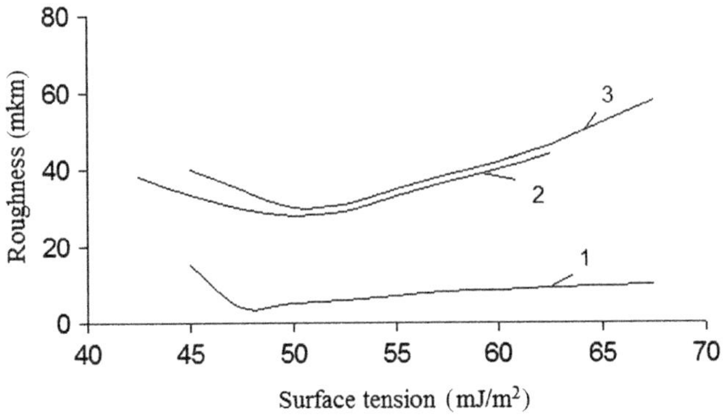

Figure 1.1 Dependence of surface roughness on surface tension colorful composition:
1—coating of PVAC
2—polymer-calcareous coating
3—calcareous coating.

For a polyvinyl acetate cement coating (PVAC) coating the equation has the form:

$$R_a = 124.8 + 3.99\sigma_{st} + 0.032\sigma_{st}^2 \quad (1.6)$$

For the calcareous coating:

$$R_a = 330 + 12.1\sigma_{st} + 0.12\sigma_{st}^2 \quad (1.7)$$

For a polymer-calcareous coating (Table 1.5):

$$R_a = 136.13 + 4.01\sigma_{st} + 0.039\sigma_{st}^2 \quad (1.8)$$

The quality of the appearance of coatings is also determined by other factors, including the degree of surface preparation of the substrate, which is characterized by an index of surface porosity. It has been established in Refs. [8,9] that with a surface porosity of more than 5%, leveling with putty in the range of 50% of the area and above is required. The different surface porosity of the cement substrate leads to a change in the water-retaining capacity of paint, and consequently, to a different quality of the appearance of the coatings.

The results of a study of the roughness of coatings (Table 1.6) on smooth glass-skimmed surface with zero porosity and on cement surfaces with different porosity confirm the assumption that the quality of the substrate and its degree of homogeneity, the presence or absence of impurities on its surface, and its porosity exert a significant influence on the quality of the appearance of the coatings formed, and thus on the protective properties of coatings.

It is obvious that an increase in porosity from 3% to 5% to 10% to 12% leads to a deterioration in the quality of the appearance of the coatings.

Table 1.5 Quality of appearance depending on rheological properties of color compositions

Type of paint	Surface tension* (mN/m)	Dynamic viscosity η^* (Pa s)	Conditional viscosity B3-246 (s)	Water retention capacity (d/dc)	Index roughness R_a (mkm)	
					Glass	Cement-sand mortar
calcareous						
1	42.79	8.12	–	3.0	41.1	50.5
2	51.96	2.92	20.0		33.9	V class
3	66.14	2.0	15.0		56.6	
Polymer-calcareous						
1	40.42	8.34	–	2.5	0.0	41.3
2	50.8	3.02	25.0		33.3	V class
3	60.63	2.33	18.0		43.5	
PVAC						
1	45.47	7.99	–	2.0	13.6	14.5
2	48.5	3.39	35.0		4.5	IV class
3	68.25	2.15	25.0		7.7	

Table 1.6 Infusion of the porosity of the substrate on the quality of the coating

Type of coating	Porosity of the substrate (%)	Roughness of coatings (mkm)	Quality class of appearance
Lime	0	33.9	V
	5.7	50.5	V
	9.6	83.4	VI
Polymer-calcareous	0	33.5	V
	4.1	41.3	V
	11.9	88.1	VI
PVAC	0	4.5	V
	3.09	14.5	IV
	10.21	46.2	V

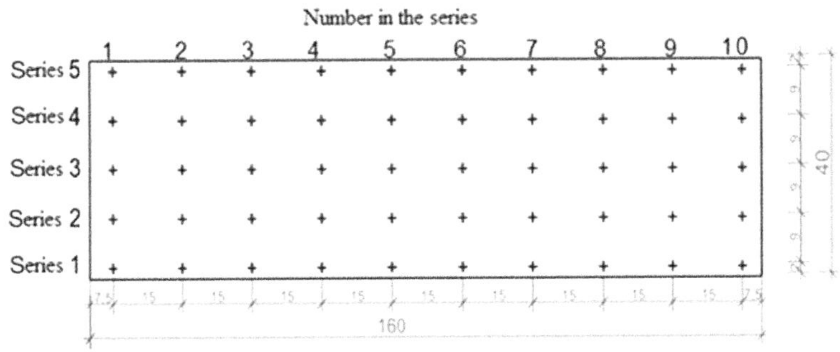

Figure 1.2 Scheme for measuring the surface roughness of paint coatings on a cement substrate (all dimensions are in mm).

1.3 Statistical analysis of the quality of appearance of paint and varnish coatings of cement concretes

To analyze the regularities of the distribution of roughness the following experiment was carried out. Paints [enamel alkyd grade PF-115, oil paint MA-15, acrylic water-dispersion paint (facade paint)] were applied by brush, pouring, and an airless method on the cement substrates with a porosity of 24%, 28%, and 32%, respectively, in two layers with intermediate drying in 20–40 minutes. There were 50 measurements on each surface by scheme (Fig. 1.2).

The results of the studies are given in Tables 1.7–1.9 and in Figs. 1.3–1.5.

Analysis of the data (Tables 1.7–1.9) indicates that the value of the roughness of the coating surface depends on the method of application of the paint composition, its rheological properties, and the porosity of the cement substrate. Thus, for oil paint MA-15 (green color), the minimum roughness value $R_a = 3.12$ mkm is achieved on a substrate with a porosity of $P = 24\%$ with a paint viscosity of

Table 1.7 Roughness of the coating surface based on PF-115, R_a, mkm

The porosity of the substrate (%)	Method of application		
	Brush	Method pouring	Airless method
24	3.13	2.97	6.37
	2.35	2.21	5.03
28	1.79	2.99	7.65
	2.23	2.37	5
32	4.34	4.31	6.97
	2.89	5.78	9.58

Note: Above the line is the average arithmetic roughness value for the viscosity of the paint 0.001×10^3 Pa s, under the line with a paint viscosity of 0.00026×10^3 Pa s.

Table 1.8 Roughness of the coating surface based on MA-15, R_a, mkm

The porosity of the substrate (%)	Method of application		
	Brush	Method pouring	Airless method
24	3.12	5.6	4.37
	6.3	4.22	5.37
28	4.26	4.04	4.53
	6.4	4.17	6.01
32	5.65	3.1	5.4
	3.87	4.1	5.21

Note: Above the line is the average arithmetic roughness value for the viscosity of the paint 0.00261×10^3 Pa s, under the line with a paint viscosity of 0.0014×10^3 Pa s.

Table 1.9 Roughness of coating based on water-dispersed paint (facade) paint, R_a, mkm

The porosity of the substrate (%)	Method of application		
	Brush	Method pouring	Airless method
24	6.5	5.54	–
	4.78	4.85	–
28	4.8	3.69	–
	4.26	4.16	–
32	3.6	4.25	–
	3.45	4.69	–

Note: Above the line is the average arithmetic roughness value for the viscosity of the paint 0.0347×10^3 Pa s, under the line with a paint viscosity of 0.013×10^3 Pa s.

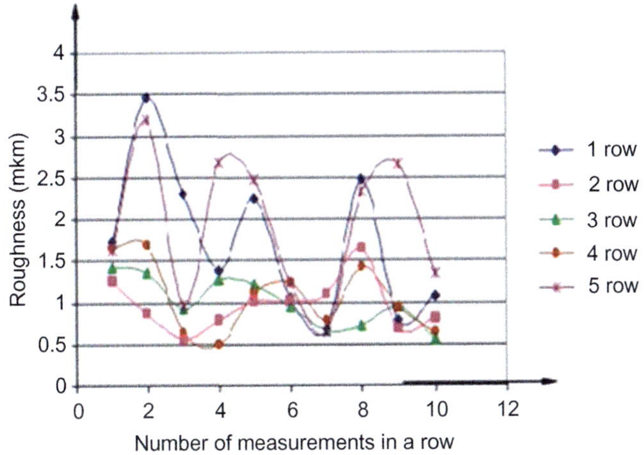

Figure 1.3 The change in the roughness of the surface of paint coatings. Paint PF-115 with a viscosity of 0.00065 Pa s, with the airless method (on substrate applied putty).

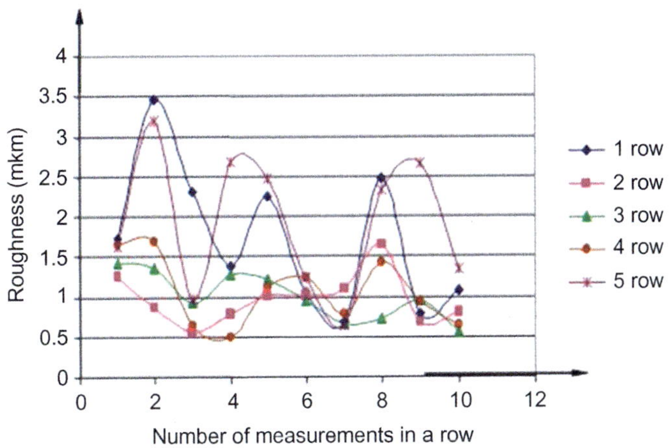

Figure 1.4 Change in the roughness of the surface of paint coatings (paint MA-15 with a viscosity of 0.0026 Pa s, applied with a brush on the substrate with a porosity of 24%).

0.00261×10^3 Pa s when applied with a brush, and with porosity $P = 32\%$ using the method pouring. For PF-115 paint, the minimum roughness value $R_a = 2.23$ mkm is achieved on a substrate with a porosity of $P = 28\%$ with a paint viscosity of 0.00026×10^3 Pa s when applied with a brush. For water-dispersion paint, the minimum roughness value $R_a = 3.45$ mkm is achieved on a substrate with a porosity of $P = 32\%$ with a paint viscosity of 0.013×10^3 Pa s when applied with a brush, and a maximum roughness value $R_a = 6.5$ mkm is achieved on a substrate with a porosity of $P = 24\%$ with an ink viscosity of 0.0347×10^3 Pa s when applied with a brush.

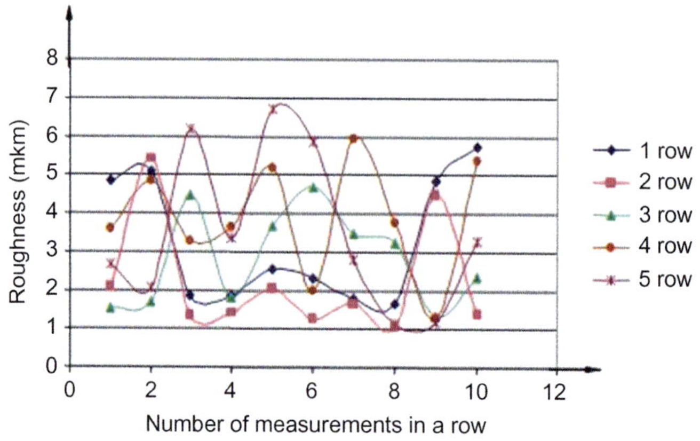

Figure 1.5 Change in the roughness of the surface of paint coatings (application of PF-115 paint by pouring on the substrate with a porosity of 28%); viscosity of the paint $\eta_1 = 0.001$ Pa s.

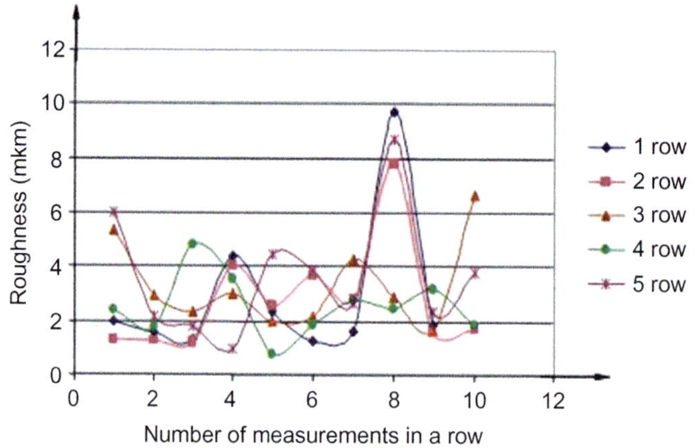

Figure 1.6 The change in the roughness of the surface of paint coatings (applying PF-115 paint by pouring on the substrate with a porosity of 28%); viscosity of the paint is $\eta_2 = 0.00065$ Pa s.

When the paint formulations are applied on substrate (putty applied), a pronounced edge effect is observed. In rows 1 and 5 a large value of the roughness index R_a is observed. On the substrates with different porosity the edge effect was not observed. This is obviously due to the predominant influence of the substrate itself and the defects of the emerging coating.

After applying paint compositions with different rheological characteristics with different methods of application on substrates with different porosity, it was not possible to detect a clear pattern of changes in the roughness depending by the above factors (Figs. 1.3–1.7).

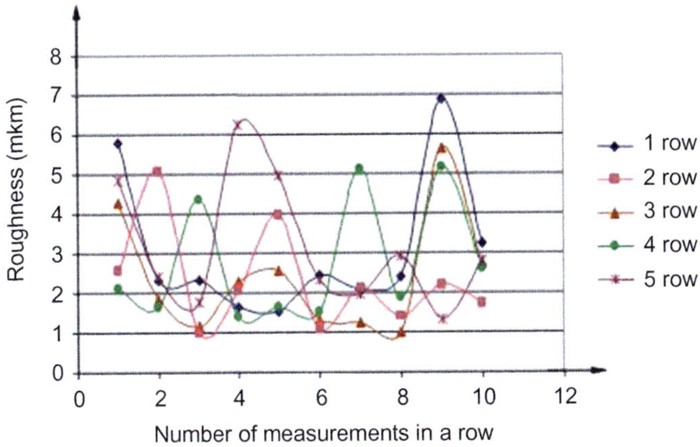

Figure 1.7 Change in the roughness of the surface of paint coatings (application of PF-115 paint on the substrate with a porosity of 28% by pouring); viscosity of the paint $\eta_3 = 0.00026$ Pa s.

To assess the uniformity of distribution of roughness parameters statistical indicators were calculated (Table 1.10).

It is established that when applying PF-115 paint on a substrate with a porosity of 24% the range of roughness values R is from 5.93 to 12.81 mkm. For MA-15 paint the range is from 5.95 to 11.33 mkm. The range between the roughness value R of the surface of coatings based on PF-115 paint on the surface (applied putty) is much lower at 2.96–4.42 mkm. For coating based on paint PF-115 with a viscosity of 0.001×10^3 Pa s characteristically more homogeneous distribution of roughness, when the paint is applied to a substrate with a porosity of 24% method pouring. The range of the roughness R_a is $R_a = 5.6$ mkm. Irrespective of the method of application, a smaller spread of the roughness value R_a is characteristic for the surfaces of all coatings on substrate (applied putty).

The results of calculating the standard deviation σ and mathematical expectation show that the roughness of the surface of coatings of different uniformity depends on the rheological properties of the paint, the porosity of the substrate, and the method of application. For paint PF-115 with a viscosity of 0.001×10^3 Pa s at porosity of the substrate at 32% a better coating is formed when the paint is applied with a brush. With this method of application a smaller spread of the roughness value is observed.

In Figs. 1.8–1.10 the histograms of the distribution of parameters of a roughness of a surface of coatings are given.

Analysis of the data (Figs. 1.8–1.10) indicates that the roughness distribution can be described by the normal distribution law. The correspondence of the empirical distribution law to the hypothetical was checked by the Pearson criterion χ^2at a significance level of 0.05.

In the statistical analysis the following points were considered: the latitude of the distribution with respect to the latitude of the tolerance field, the center of the

Table 1.10 Statistical indices of the processing of sample data

Type of the paint	The porosity of the substrate (%)	Standard deviation (σ)/the range of data (R)								
		Method for applying paint composition								
		Brush			Pouring			Airless		
		Viscosity (Pa s × 10^3)			Viscosity (Pa s × 10^3)			Viscosity (Pa s × 10^3)		
		η_1	η_2	η_3	η_1	η_2	η_3	η_1	η_2	η_3
PF-115	24	4.11	2.71	1.5	1.95	1.63	1.48	2.88	2.8	3.02
		12.81	10.55	5.93	7.33	5.6	5.9	10.94	9.81	12.25
	28	0.86	0.64	1.69	1.63	1.93	1.52	1.92	2.38	2.83
		3.96	2.54	7.3	5.64	8.96	5.89	8.61	8.8	12.23
	32	1.74	2.14	1.8	1.97	2.5	2.22	3.32	2.24	2.5
		6.96	9.3	6.37	7.66	7.95	8.05	11.21	8.66	8.41
	putty	0.43	0.71	0.97	0.53	0.71	0.31	0.95	0.71	0.83
		1.8	2.86	4	2.25	3.4	1.64	3.68	2.96	3.74
MA-15	24	1.65	2.51	1.56	1.6	1.8	1.7	2.94	1.85	2.73
		5.95	11.33	6.97	6.9	6.33	6.31	10.12	6.65	10.76
	28	1.23	1.99	1.56	1.86	1.95	1.91	2.6	2.48	2.91
		4.34	8.38	6.17	7.41	8	6.92	10.31	9.96	11
	32	2.12	1.61	1.44	1.27	1.43	1.79	5.61	2.45	2.65
		8.12	8.37	5.22	5.37	4.94	7.21	10.1	9.35	9.81
	putty	0.78	0.93	0.87	1.19	1.02	1.4	1.21	0.89	0.96
		2.96	4.42	4	3.68	3.62	4.68	4.52	3.35	4.06
Water-dispersed	24	2.15	1.62	2.18	2.52	1.86	2.66	–	–	–
		7.45	5.52	6.97	10.63	7.85	10.24	–	–	–
	28	1.81	2.01	1.31	2.17	1.75	2.2	–	–	–
		7.78	8.67	5.37	7.7	8.34	7.63	–	–	–
	32	2.11	0.96	1.3	2.63	1.06	2.55	–	–	–
		9.56	3.58	5.2	10.39	5.08	8.95	–	–	–
	PS.	1.02	0.96	0.95	1.05	2.52	1.52	–	–	–
		4.46	3.46	3.7	4.1	8.92	6.81	–	–	–

Note: PS is putty substrate. For PF-115 paint, $\eta_1 = 0.001$ Pa s, $\eta_2 = 0.00065$ Pa s, $\eta_3 = 0.00026$ Pa s; for paint MA-115 $\eta_1 = 0.0026$ Pa s, $\eta_2 = 0.002$ Pa s, $\eta_3 = 0.0014$ Pa s; for water-dispersion paint $\eta_1 = 0.0347$ Pa s, $\eta_2 = 0.02317$ Pa s, $\eta_3 = 0.013$ Pa s.

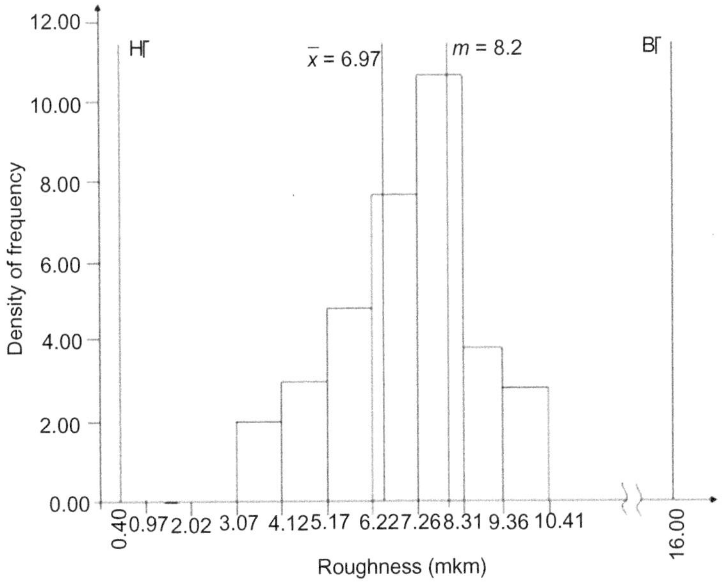

Figure 1.8 The histograms of the distribution frequencies of the surface roughness on the basis of PF-115 paint applied using the airless method (paint viscosity 0.001×10^3 Pa s; porosity of the substrate 32%).

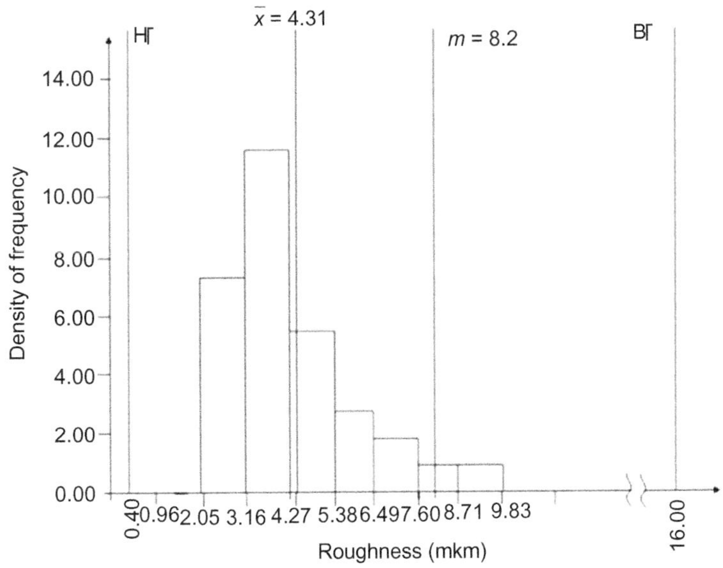

Figure 1.9 The histograms of the distribution frequencies of the surface roughness of the coating based on PF-115 paint, applied by pouring (viscosity of paint 0.001×10^3 Pa s; porosity of the substrate 32%).

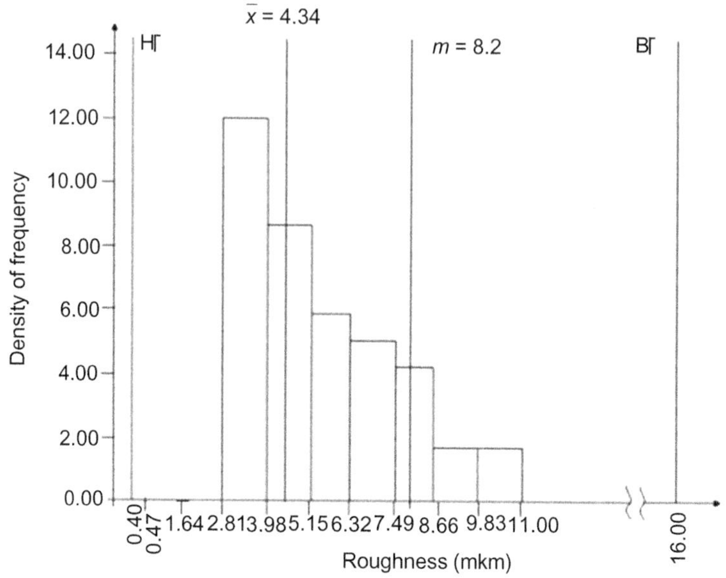

Figure 1.10 The histograms of the distribution frequencies of the surface roughness of the coating based on PF-115 paint, applied by a brush (paint viscosity 0.001×10^3 Pa s; porosity of the substrate 32%).

distribution with respect to the center of the tolerance field, and the form of the distribution [10–13].

As seen in Fig. 1.8 the shape of the distribution is satisfactory, the distribution center is equal to 6.97 μm, the center of the tolerance field is equal to $x = 8.2$ mkm, and the latitude of the distribution is approximately 3/5 of the tolerance field. Consequently, the quality of the products, in this case the quality of the appearance of the coatings, is satisfactory.

In Fig. 1.9 the distribution center is displaced. The histogram is shifted to the left because of its asymmetry (the average value is 4.31); the distribution center does not coincide with the center of the tolerance field, $x = 7.8$ mkm; and the latitude of the distribution is 1/2 of the width of the tolerance field. All this indicates a defective surface.

In Fig. 1.10, the distribution center (equal to 4.34 mkm) is shifted to the lower tolerance limit of 0.4 mkm, and does not coincide with the center of the tolerance field. The left side of the distribution at the boundary with the lower tolerance is a steep bank. This means that the technological process of obtaining coatings with specified properties has been violated.

Any production process for the manufacture of products can include one-sided and two-sided tolerances with specified values (UTL and LTL refer to upper and lower tolerance limits). To effectively analyze the process, indices of reproducibility can be used that determine whether the process has sufficiently low variability and satisfies the process tolerances or if there is a tuning problem.

We used the reproducibility index C_p to evaluate the working capacity of the coating process with a given appearance quality when applying paint formulations to a substrate in various ways. The reproducibility index C_p was calculated by the formula:

$$C_{pk} = \min\left\{\frac{\overline{x} - \text{LTL}}{3\sigma}; \frac{\text{UTL} - \overline{x}}{3\sigma}\right. \tag{1.9}$$

where UTL is the upper tolerance limit and the LTL is the lower tolerance limit. $\sigma$is the standard deviation.

In accordance with ISO 1302, there are 12 surface roughness classes for which tolerances are defined. Table 1.11 shows the numerical values of index C_{pk} for the process of obtaining coatings based on PF-115 ink with a specified appearance quality, characterized by N5.

Analysis of the data (Table 1.11) indicates low reproducibility of the process of producing a coating characterized by class N5. The process is reproducible with a given quality of appearance based on acryl water-dispersed (facade) paint on the substrate with a porosity of 32% and a viscosity $\eta = 0.02371 \times 10^3$ Pa s when applied with a brush and pouring. The process is reproducible on the surface of the putty when the viscosity of the paint $\eta = 0.013 \times 10^3$ and 0.02371×10^3 Pa s when applied with a brush and paint with viscosity $\eta = 0.0347 \times 10^3$ Pa s with pouring. The process is reproducible at obtaining coatings on the substrate with a porosity of 24% with viscosity of paint $\eta = 0.0347 \times 10^3$ Pa s and $\eta = 0.02371 \times 10^3$ Pa s when applied with a brush. For other methods of dyeing process is no reproducible. The process is reproducible when paint PF-115 is used with viscosity $\eta = 0.00026 \times 10^3$ Pa s using the airless method on a substrate with a porosity of 32%.

On the surface of the putty process is reproducible, with viscosity of paint $\eta = 0.001 \times 10^3$ Pa s when applied by brush and airless method, as well as with a viscosity of paint $\eta = 0.00026 \times 10^3$ Pa s when applied of method pouring. The process is reproducible for coatings based on oil paint MA-15 (N5 class), when the viscosity of paint $\eta = 0.0026 \times 10^3$ Pa s when applied with a brush on a substrate with a porosity of 28%, and on a substrate with a porosity of 24% using the pouring method. The process is reproducible when the viscosity of a paint MA-15 $\eta = 0.0014 \times 10^3$ Pa s and application a brush onto the substrate with a porosity of 24% and 28%. The process is reproducible, when the viscosity of paint 0.0020×10^3 Pa s at a substrate staining with a porosity of 28% and application a brush and porosity 32% application of method pouring. The process is reproducible if the porosity of the substrate is 28%, the viscosity of paint $\eta = 0.0020 \times 10^3$ Pa s with dyeing substrate with a porosity of 28% and 32% brush using the pouring method. The process is reproducible if viscosity of paint $\eta = 0.0020 \times 10^3$ Pa s and $\eta = 0.0014 \times 10^3$ Pa s when applying using a brush and the airless method.

Table 1.11 Values of the index C_{pk} for the process of producing coatings

Type of the paint	Porosity the cement substrate (%)	Values of the index C_{pk}								
		A method of applying								
		Brush			Method pouring			Airless method		
		Viscosity (Pa s × 10^3)			Viscosity (Pa s × 10^3)			Viscosity (Pa s × 10^3)		
		η_1	η_2	η_3	η_1	η_2	η_3	η_1	η_2	η_3
PF-115	24	0.272	0.428	0.581	0.548	0.739	0.559	0.745	0.578	0.570
	28	0.803	**1.303**	0.493	0.659	0.552	0.578	**1.329**	**1.049**	0.605
	32	0.863	0.440	0.579	0.759	0.537	0.883	0.705	0.81	**1.267**
	putty	**1.23**	0.986	0.715	0.873	0.719	**2.005**	**1.001**	0.706	0.825
MA-15	24	0.678	0.992	**1.363**	**1.189**	0.758	0.864	0.514	0.972	0.670
	28	**1.203**	**1.062**	**1.380**	0.758	0.761	0.759	0.601	1.215	0.698
	32	0.492	0.819	0.940	0.866	**1.138**	0.795	0.328	0.883	0.671
	putty	0.886	**1.002**	**1.049**	0.788	0.902	0.793	0.852	**1.283**	**1.060**
Acryl water dispersed (façade) paint	24	**1.020**	**1.000**	0.754	0.747	0.653	0.453	–	–	–
	28	0.910	0.596	1.127	0.599	0.725	0.657	–	–	–
	32	0.595	**1.377**	0.939	0.560	**1.289**	0.633	–	–	–
	putty	0.926	**1.149**	**1.355**	**1.147**	0.406	0.916	–	–	–

Note: 1. At $C_{pk} > 1.33$—the process is reproducible; $1 < C_{pk} < 1.33$—the process requires careful attention; $C_{pk} < 1$—the process is not reproducible. 2. PF-115, viscosity is $\eta_1 = 0.001 \times 10^3$ Pa s; paint PF-115, viscosity is $\eta_2 = 0.00065 \times 10^3$ Pa s; paint PF-115, viscosity is $\eta_3 = 0.00026 \times 10^3$ Pa s. 3. Paint MA-15, viscosity is $\eta_1 = 0.0026 \times 10^3$ Pa s; paint MA-15, viscosity is $\eta_2 = 0.0020 \times 10^3$ Pa s; paints MA-15, viscosity is $\eta_3 = 0.0014 \times 10^3$ Pa s. 4. Acryl water dispersed (façade) paint, viscosity is $\eta_1 = 0.0347 \times 10^3$ Pa s; acryl water dispersed (façade) paint, viscosity is $\eta_2 = 0.02371 \times 10^3$ Pa s; acryl water dispersed (façade) paint, viscosity is $\eta_3 = 0.013 \times 10^3$ Pa s.

Thus, the above-mentioned results of the studies and calculations convincingly show the low reproducibility of the process of obtaining the painted surface of building products and structures with a given quality. This is one of the reasons for the discrepancy between the actual service life and the forecasted life of these products. When evaluating the reproducibility of the coating process with a different quality of appearance (e.g., with class N6), the processes of staining of cement substrates with different porosity and other rheological characteristics may also be reproducible. However, undoubtedly, the use of statistical analysis makes it possible to develop recommendations on the choice of the method and the staining regime, which will help to improve the quality of paint and varnish coatings and increase the period of overhauls.

1.4 Patterns of quality change of the appearance of paint coatings of cement concretes with aging

Coatings for exterior decoration of buildings should perform not only protective functions, but also maintain high decorative properties during operation. The main types of decorative properties of coatings are shine, color, mud retention, and chalking. Weathering, cracking, peeling, bubbling, and the appearance of corrosion foci on the coating are used to characterize the protective properties of coatings. Change in shine is the first sign of the beginning of the destruction of the coating. It is caused by the destruction of the surface lacquer layer, leading to an increase in the size and height of the particles and aggregates of pigments protruding on the surface of the coating.

As the surface layer of the film former disintegrates, the aggregates and pigment particles are gradually exposed on the surface of the coating, indicating the initial stage of chalking of the coatings [14,15]. According to Ref. [16], the time to the initial stage of chalking coincides with the time necessary to reduce the shine by 50% of the initial.

In natural conditions, polymer coatings are exposed to the relative humidity of the air (rain, snow, dew, and fog). The action of moisture in the form of rain is relatively short-lived, and the rainwater, penetrating into the film, evaporates rather quickly, without causing serious damage. Heavy rains can wash out water-soluble compounds, wash the chalking layer, and cause mechanical damage to the coatings. More dangerous is the effect of relative air humidity and dew, as during the formation of dew, moisture condenses in the pores of the coatings.

As a result of the relative humidity of the air, moisture sorption takes place. Its magnitude is determined by the relative humidity of the air and the properties of the coatings. For coating, that contain polar groups, at a high relative humidity of the air, multilayer adsorption or capillary condensation occurs. In the range of values of relative air humidity from 0% to 60%–80%, the dependence of sorption on the relative air humidity is close to linear. Further, starting from 60% to 80%, a sharp increase in the sorption of moisture occurs [17].

The sorption of moisture depends on the nature of the coating, its polarity, packing density, the density of the spatial grid, the presence of plasticizer and various additives, nature, and the volume concentration of pigments and fillers.

When exposed to moisture on the coatings, physical, chemical and photochemical processes occur. As a result of physical processes, moisture absorption can lead to plasticization of the film-forming agents and an increase in the gloss of the coatings.

Fig. 1.11 shows the changes in the shine of organosilicon KO-168 coatings as a function of the relative humidity of air under ultraviolet irradiation with a PRK-2 lamp at an intensity equal to 57 W/m^2. At the same exposure time of the coatings, the dependence of the shine on the relative humidity of the air is described by the exponential function

$$y = A\exp(\alpha\phi) \tag{1.10}$$

After calculating the coefficients A, the equation has the form

$$S = 793\exp(-0.00\varphi) - \text{for organosilicon KO-168 coatings} \tag{1.11}$$

$$S = 55.8\exp(0.0012\varphi) - \text{for PVACs} \tag{1.12}$$

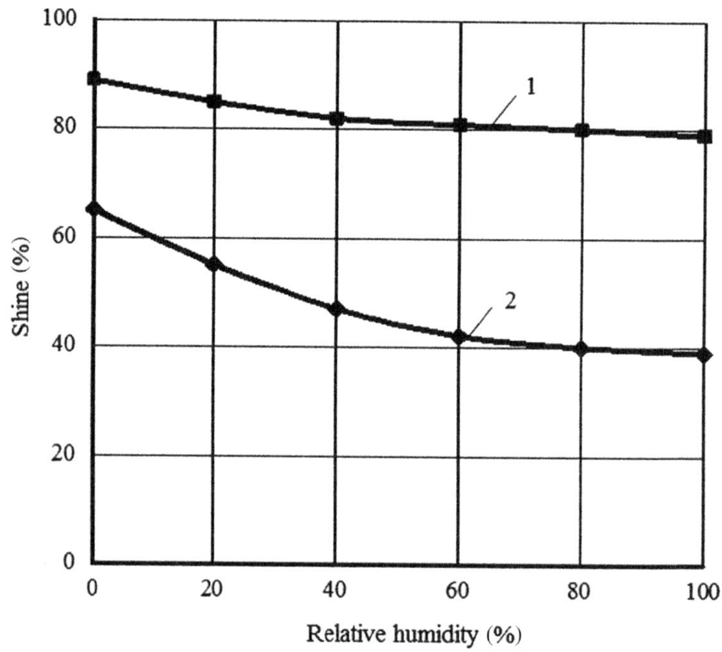

Figure 1.11 Dependence of the shine of polymer coatings on the relative humidity of air
1—coating of KO-168
2—coating of PVAC.

Analysis of the experimental data (Fig. 1.11) shows that the change in the shine of the coatings in time is described by an equation of the form

$$S = A\exp(\alpha\tau) \tag{1.13}$$

The method of least squares was used to calculate the values of the coefficients A and α. For organosilicon coatings on a glass dense substrate, the dependence of the shine on exposure time in an environment with a moisture content of 60% has the form

$$S = 87\exp(-0.004\tau) \tag{1.14}$$

and in an environment with a humidity of 100%

$$S = 87\exp(-0.005\tau) \tag{1.15}$$

Comparative analysis of the obtained dependences shows that with increasing relative air humidity, the value of the coefficient increases, that is, the degree of influence of moisture on the variability of coatings increases.

At hardening of the silicone coatings on a cement substrate with a porosity of $P = 20.3\%$, the dependence of the shine on the exposure time has the form

$$S = 52.48\exp(-0.0072\tau) \quad \text{for air humidity 60\%} \tag{1.16}$$

$$S = 52.48\exp(-0.0093\tau) \quad \text{for air humidity 100\%} \tag{1.17}$$

Thus, the degree of influence of the relative humidity of the air on the change in the shine of the coatings on a porous substrate affects shine to a greater extent than coatings on a dense glass substrate. Increasing the porosity of the cement substrate contributes to a greater dependence of the shine of the coatings on the relative humidity of the air. For organosilicon KO-168 coatings on a cement substrate with porosity $P = 26\%$, the equation has the (Fig. 1.12).

$$S = 52.48\exp(-0.0072\tau) \quad \text{for air humidity 60\%} \tag{1.18}$$

$$S = 52.48\exp(-0.0093\tau) \quad \text{for air humidity 100\%} \tag{1.19}$$

For PVACs the equation has the form

$$S = 62\exp(-0.017\tau) \quad \text{for air humidity 60\%} \tag{1.20}$$

$$S = 62\exp(-0.02\tau) \quad \text{for air humidity 100\%} \tag{1.21}$$

It was found that at the initial moment of moistening an increase in shine is observed, which is apparently due to the plasticizing action of moisture, which

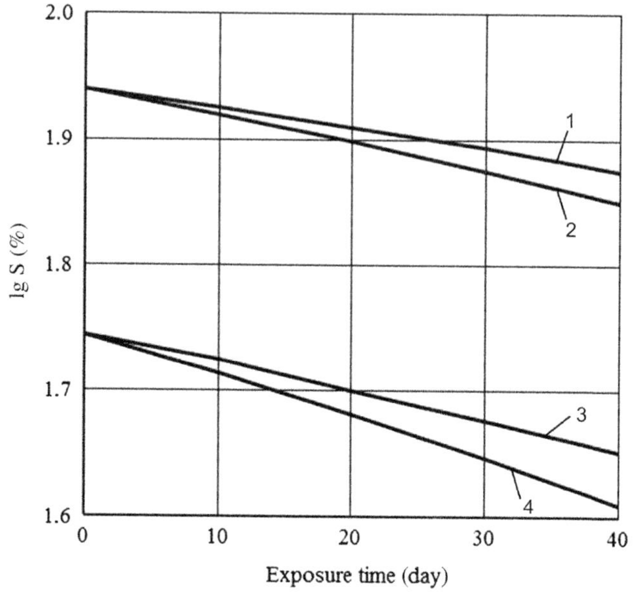

Figure 1.12 The dependence of the shine of coatings on exposure time
1—coating of KO-168, air humidity $\varphi = 60\%$
2—coating of KO-168, air humidity $\varphi = 100\%$
3—coating of PVAC, air humidity $\varphi = 60\%$
4—coating of PVAC, air humidity $\varphi = 100\%$.

facilitates relaxation processes. Spontaneous leveling of the surface microrelief also occurs (Fig. 1.14). The change in the surface area of the coating was evaluated from the change in the adsorption of dyes from solutions. A change in the intensity of the color of the solutions after contacting them with the surface of the coating was determined. A solution of methylene blue dye in acetone was used. As an example, we considered the moistening process as one of the special cases of aging.

Analysis of the data (Fig. 1.13) shows that at the first moment of moistening, an increase in the shine of the coatings is observed due to the plasticizing effect of moisture. This leads to an equalization of the microrelief of the coating surface. This is evidenced by data on the change in the adsorption area of the dye (curve 3). After 800 h of moistening, an increase in the surface area of the coating is observed. After 800 hours of moistening, the initial stage of chalking is observed. Thus, after 800 hours of moistening, the degree of chalking of the coating is estimated at 1 point, after 1000 hours at 2 points, and after 1200 hours at 3 points.

Under the action of UV irradiation in the initial aging period, intensive destruction of the coatings is observed, then the rate of destruction gradually decreases. For AK-111 coatings after 120 hours of aging, the weight reduction m is 55%, for coatings PVAC it is 45%.

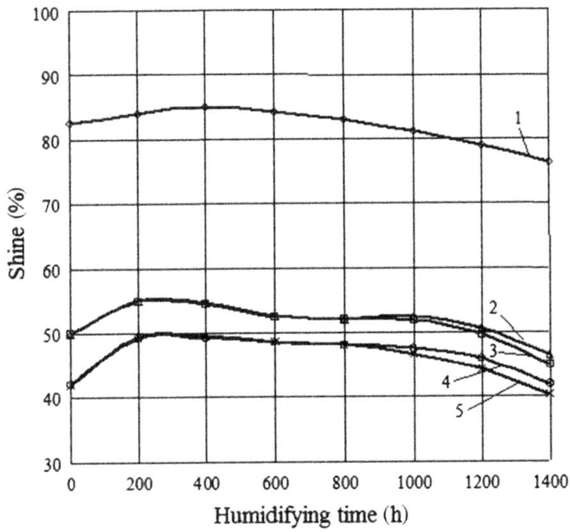

Figure 1.13 Change in the shine of coatings during wetting:
1—coating of KO-168
2—coating of VD-AK-111 at a temperature of 20°C
3—coating of VD-AK-111 at a temperature of 60°C
4—coating of PVAC at a temperature of 20°C
5—coating of PVAC at a temperature of 60°C.

It was found that at UV irradiation the initial stage of chalking of the coatings is manifested after 100 hours of UV irradiation. The degree of chalking of coatings after 100 hours of exposure is estimated at 2 points, along with an increase in the area of coatings is observed (Fig. 1.14). There is a change in the surface area of the coatings, which causes a decrease in the quality of their appearance. The intensity of destruction under the influence of climatic factors is not the same at different stages of operation. It was found that during some incubation period (the duration of which depends on the type of coating), there is a slight increase, and possibly a decrease in defectiveness. Then, in the active stage of accumulation of damages, significant changes occur in the structure and properties of the coating (change in color, shine, cracking, etc.), which ultimately leads to failure of the coating.

Figs. 1.15–1.18 shows the results of measuring the surface roughness of coatings exposed to moisture.

Analysis of the obtained data allows us to state, that, regardless of the rheological properties of paint and the porosity of the substrates, the surface roughness is reduced during the moistening of the coatings in the first stage (up to 30 days), that is, the surface microrelief is leveled due to the plasticizing effect of moisture (swelling of the coatings). In the future, due to the destructive effect of moisture, the surface roughness increases, caused by the appearance of microcracks, rashes, and bubbles.

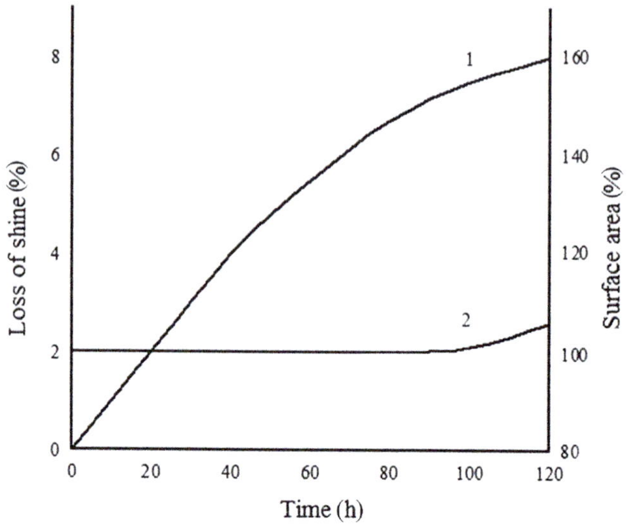

Figure 1.14 Change in the properties of PVAC coatings in the process of UV irradiation:
1—loss of shine
2—surface area.

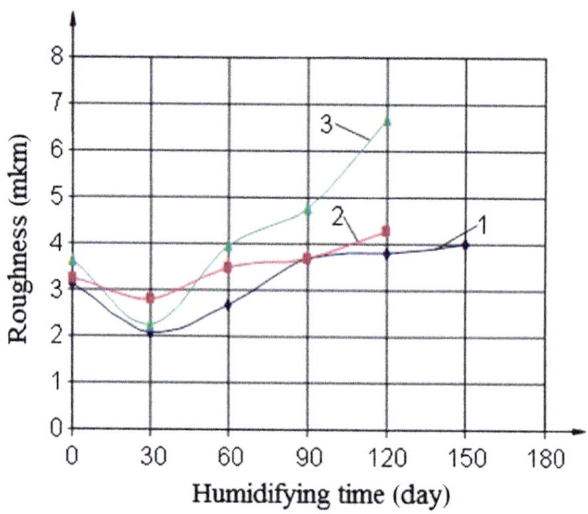

Figure 1.15 Change in the roughness of coatings based on PF-115 in the process of moistening (the paint was applied by brush; the porosity of the substrate was 24%):
1—viscosity of the paint 0.001×10^3 Pa s
2—viscosity of paint 0.00065×10^3 Pa s
3—viscosity of the paint 0.00026×10^3 Pa s.

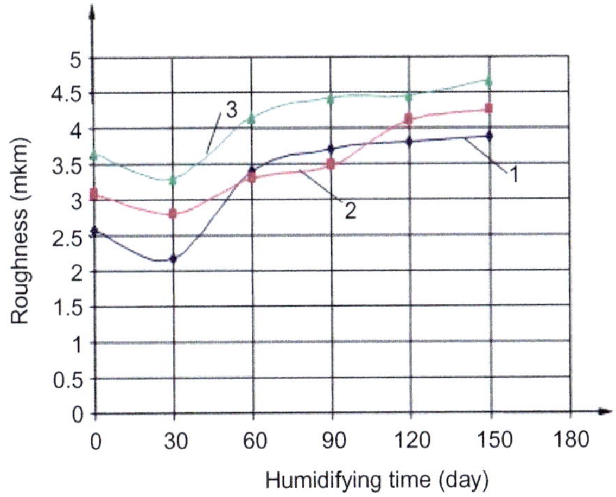

Figure 1.16 Change in surface roughness of coatings based on water-dispersion paint in the process of moistening (paint was applied with a brush, putty-coated substrate):
1—viscosity of the paint 0.0347×10^3 Pa s
2—viscosity of the paint 0.02317×10^3 Pa s
3—viscosity of the paint 0.013×10^3 Pa s.

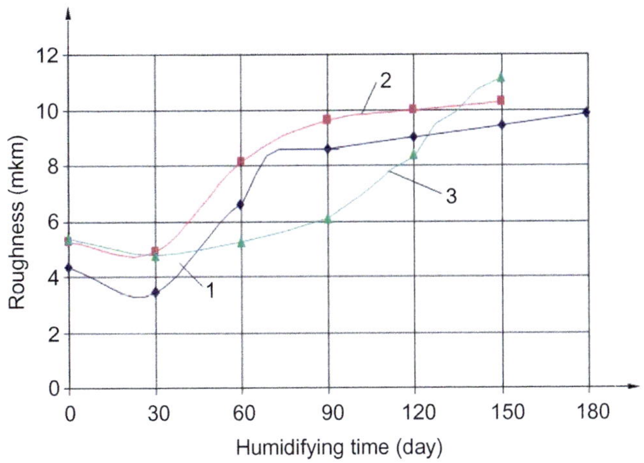

Figure 1.17 Change in the roughness of coatings based on MA-15 paint in the process of moistening (paint pneumatic dispersion, substrate with a porosity of 24%):
1—the viscosity of the paint 0.0026×10^3 Pa s
2—viscosity of the paint 0.002×10^3 Pa s
3—viscosity of the paint 0.0014×10^3 Pa s.

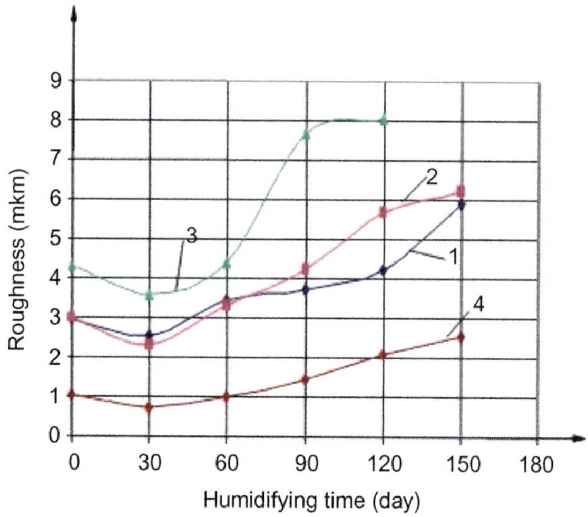

Figure 1.18 Change of roughness of coatings in the process of moistening (application method pouring):
1—paint PF-115 with a viscosity of 0.001×10^3 Pa s, and porosity of the substrate is 24%
2—paint PF-115 with a viscosity of 0.001×10^3 Pa s, and porosity of the substrate 28%
3—PF-115 paint with a viscosity of 0.001×10^3 Pa s, and porosity of the substrate is 32%
4—paint PF-115 viscosity of 0.001×10^3 Pa s, and the surface of the substrate is plastered.

For coatings based on water-dispersion paint the decrease in the viscosity of the paint composition contributes to an increase in the roughness of the coating surface (Fig. 1.16). Thus, with a paint viscosity of 0.0347×10^3 Pa s, the roughness value of the coating surface was $R_a = 2.5$ mkm, and with a viscosity of 0.013×10^3 Pa s, 3.7 mkm. Similar patterns are also seen for coatings based on MA-15 oil paint (Fig. 1.17).

Analysis of the data (Fig. 1.18) shows that when the porosity of the substrate is increased, increased roughness of the surface of coatings based on PF-115 paint applied with the same viscosity of 0.001×10^3 Pa s is observed, which is obviously due to the influence of substrate structure. Thus, with a substrate porosity of 32%, the initial roughness of the coating surface is $R_a = 4.3$ mkm, and after 120 days of humidification, it is 8 mkm, while on a substrate with a porosity of 24%, it is 3 and 6 μm. Application of putty on surface of the substrate significantly reduces the surface roughness of the coating (R_a is $R_a = 1$ mkm before humidification, 2.6 mkm after humidification), and a lower rate of degradation is observed.

The data obtained (Figs. 1.15–1.19) correlate well with other indicators of the protective and decorative properties of the coating (color change, shine variation, chalking, mud retention, bronzing, etc.). After 150 days of humidification, the shine of coatings based on PF-115 paint is decreased by 6.7%–25.5%, based on MA-15 paint it is 2.7%–15.4%, and based on water-dispersion paint it is 3.5%–6.53%. Thus, the shine of coatings based on PF-115 ink with a viscosity of

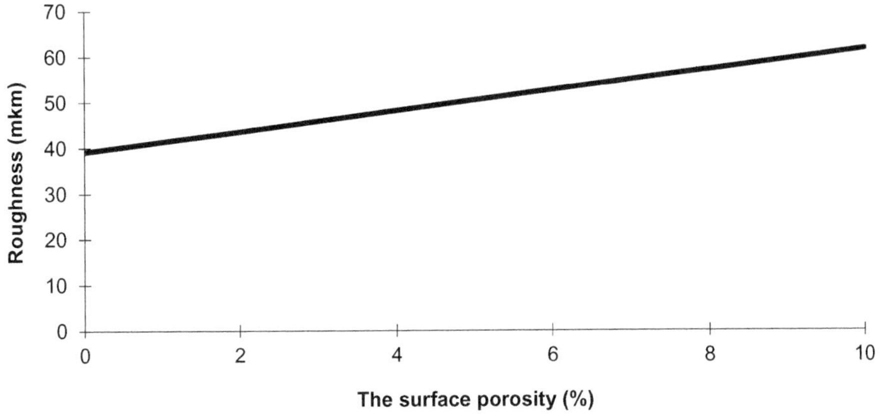

Figure 1.19 Effect of the surface porosity of the cement substrate on the quality of the appearance of the polymer-calcareous coating.

0.001×10^3 Pa s, applied by brush on a substrate with a porosity of 24%, decreased by 16%, bubbles, rash and flaking of the paint appeared, and for paint-based coatings with a viscosity of 0.00026×10^3 and a decrease in gloss of 19%, there were peeling and rashes. Similar patterns are observed in coatings based on water-dispersion and oil paints. A decrease in shine by 4.8% and crack formation, chalking, and mud retention are observed in coatings based on water-dispersion paint with a viscosity of 0.0347×10^3 Pa s on a substrate(application of putty). For coatings based on MA-15 paint with a viscosity of 0.002×10^3 Pa s on the substrate with a porosity of 24%, the gloss reduction was 7.5%, and cracking and peeling of the coatings are observed (Tables 1.12–1.14).

In the course of the experiment it was revealed that the nature of the destruction of coatings in the aging process is not the same for all coatings. Thus, coatings based on oil and alkyd paint are characterized by peeling and the appearance of rashes and bubbles, and for coatings based on water-dispersion paint cracking is seen.

Thus, depending on the quality of the surface to be painted (porosity), the method of application, optimal rheological characteristics of the paint should be chosen to create a coating surface with the specified quality.

The results show that an increase in the surface roughness of a coating indicates a decrease in the thickness of the coating. We have calculated the change in the thickness of the coating during the corrosive action of the environment for some coatings. The results of the designs are given in Table 1.15.

Mathematical processing of the data (Table 1.15) shows that the dependence of the change in coating thickness during the corrosive action of the environment can be approximated by an expression of the form

$$\Delta h = At^b \tag{1.22}$$

Table 1.12 Indicators of protective and decorative properties of coatings (the paint composition was applied by brush)

Type of destruction																						
η-P	Shine change, %, point			Color change point			Whiteness, point			Hold dirt point			Cracking, point			Peeling, point			Bubble, point		
	PF-115	MA-15	WD	PF-115	MA-15	WD	PF-115	MA-15	WD	PF-115	MA-15	WD	PF-115	MA-15	WD	PF-115	MA-15	WD	PF-115	MA-15	WD
1–24	16–4			4	4	–	4	5	–	5	4	–	6	3	–	5	4	–	4	5	–
2–24	20–4	7.5–5		4	4	–	4	5	–	5	5	–	7	3	–	5	4	–	4	5	–
3–24	19–4	2.7–5		4	4	–	4	4	–	5	4	–	7	4	–	5	5	–	4	5	–
1–28	17–4	9–4	4.5–5	4	4	4	5	5	4	5	4	5	7	3	7	5	4	5	3	5	5
2–28		7.5–5		–	4	–	–	–	–	–	4	–	–	3	–	–	4	–	–	5	–
3–28		13–4		–	–	–	–	–	–	–	–	–	–	–	–	–	–	–	–	–	–
1–32	17–4	3.3–5	4.9–5	4	4	3	3	5	3	4	5	3	7	5	7	5	5	4	3	5	5
2–32	17.8–4	a	5–5	4	4	4	5	5	4	4	3	5	7	7	7	5	5	5	4	5	5
3–32	10–4	a	4.5–4	4	4	4	5	5	4	4	3	5	7	7	7	5	5	5	4	5	5
1–0	8.5–4	4–5	3.5–5	4	4	4	5	5	5	5	4	5	5	4	7	3	4	5	5	5	5
2–0		6.77–4	4.8–5	4	4	4	4	5	5	5	4	5	7	4	7	5	4	5	5	5	5
3–0		5.5–4	7–4	4	4	4	4	5	5	5	4	5	7	4	7	5	4	5	5	5	5

[a]Bronze 3 points.

Table 1.13 Indicators of protective and decorative properties of coatings (paint was applied using the pouring method)

Type of destruction																					
η-P	Shine change, %, point			Color change point			Whiteness, point			Hold dirt, point			Cracking, point			Peeling, point			Bubble, point		
	PF-115	MA-15	WD	PF-115	MA-15	WD	PF-115	MA-15	VD	PF-115	MA-15	WD	PF-115	MA-15	WD	PF-115	MA-15	WD	PF-115	MA-15	WD
1–24	13.5–4			4	4	–	5	4	–	4	5	–	7	4	–	5	4	–	4	5	–
2–24	17.7–4			4	4	–	5	5	–	4	5	–	7	4	–	5	4	–	4	5	–
3–24	25.5–3	5.1–4		4	4	–	3	5	–	3	5	–	6	4	–	5	4	–	4	5	–
1–28	8.9–5	4.9–4	3.7–5	4	4	4	5	5	5	4	4	5	7	3	7	5	5	5	4	5	5
2–28		15.4–4		–	4	–	–	4	–	–	4	–	–	3	–	–	5	–	–	5	–
3–28				–	–	–	–	–	–	–	–	–	–	–	–	–	–	–	–	–	–
1–32		4.9–4	4.5–5	4	3	3	5	3	3	4	3	3	7	7	7	7	4	4	4	5	5
2–32	6.7–4			4	3	4	5	3	4	4	3	4	7	7	7	7	5	5		5	5
3–32	19–4		6.53–4	4	3	4	5	3	4	4	3	4	7	7	7	7	5	5	4	5	5
1–0	7.8–4			4	4	4	5	5	5	5	5	5	7	3	5	7	5	5	5	5	5
2–0		3.7–5	4.65–4	4	4	4	5	5	5	4	4	5	7	4	5	7	4	5	4	5	5
3–0	11.4–4	10.8–4	4.1–4	4	4	4	5	5	5	4	4	5	7	3	7	7	7	4	4	5	5

Table 1.14 Indicators of protective and decorative properties of coatings (paint was applied using the airless method)

Type of destruction																					
η-P	Shine change, %, point			Color change, point			Whiteness, point			Hold dirt, point			Cracking, point			Peeling, point			Bubble, point		
	PF-115	MA-15	WD	PF-115	MA-15	WD	PF-115	MA-15	WD	PF-115	MA-15	WD	PF-115	MA-15	WD	PF-115	MA-15	WD	PF-115	MA-15	WD
1–24	18.3–4	9.4–4	–	4	4	–	5	5	–	5	5	–	7	6	–	5	5	–	5	5	–
2–24	17.8–4		–	4	4	–	5	5	–	5	5	–	7	6	–	5	5	–	5	5	–
3–24			–	4	4	–	5	5	–	5	5	–	7	6	–	5	5	–	4	5	–
1–28																					
2–28	21.45–3	5.7–4	–	4	4	–	5	5	–	5	5	–	7	7	–	5	5	–	4	4	–
3–28		4.26–5	–	4	4	–	5	5	–	5	5	–	7	6	–	5	4	–	4	5	–
1–32			–	4	4	–	5	4	–	5	4	–	7	7	–	5	5	–	4	4	–
2–32																					
3–32	22.8–3	10.53–4	–	4	4	–	5	4	–	5	4	–	7	5	–	5	5	–	3	5	–
1–0	–		–	–	4	–	–	4	–	–	4	–	–	5	–	–	4	–	–	5	–
2–0		8.3–4	–	4	4	–	5	5	–	5	5	–	6	5	–	5	5	–	4	5	–
3–0	16.4–4		–	4	4	–	5	5	–	5	5	–	6	5	–	5	5	–	4	5	–

Table 1.15 Change in the thickness of the coating during the corrosive action of the environment

Name of paint	Change in thickness Δh after test cycles					
	After curing	**5 Cycles**	**8 Cycles**	**11 Cycles**	**13 Cycles**	**15 Cycles**
Alkyd Enamel PF-115	0	0.25	A peeling of the coating			
	0	0.18	A peeling of the coating			
	0	0.11	A peeling of the coating			
Oil paint MA-15	0	0.63	A peeling of the coating			
	0	0.25	A peeling of the coating			
	0	0.19	0.47	A peeling of the coating		
Nitrocellulose NC-123	0	0.24	A cracking of the coating			
	0	0.19	0.52	A cracking of the coating		
	0	0.12	0.36	0.86	1.00	Cracking
Waterdispersive (facade)	0	0.41	0.57	0.71	0.93	1.15
	0	0.21	0.41	0.61	0.73	0.79
	0	0.09	0.23	0.49	0.60	0.76
Acrylate class "Universal"	0	0.22	0.72	0.88	0.98	1.12
	0	0.17	0.49	0.66	0.87	1.06
	0	0.12	0.33	0.45	0.67	0.8

where b is an indicator characterizing the degree of aggressiveness of the environment, t refers to the test cycles, and Δh indicates the change in coating thickness.

For coatings based on the PF-115 paint, the model (1.22) has the form

$$\Delta h = 0.49t^{1.01}$$

For coatings based on the MA-15 paint

$$\Delta h = 0.13t^{0.97}$$

For coatings based on the NZ-123 paint

$$\Delta h = 0.008t^{1.87}$$

For coatings based on acrylate paint class "universal"

$$\Delta h = 0.07t^{1.00}$$

For coatings based on water-dispersion (facade) paint

$$\Delta h = 0.05t^{1.14}$$

The presented models allow estimating the degree of destruction of coatings in the process of corrosion. To a large extent the degree of fracture of coatings with a large initial roughness. Thus, after 15 test cycles, reduction in the thickness of coatings based on water-dispersion facade paint is 1.15 mkm with an initial roughness $R_a = 3.01$, while a decrease in the thickness of coatings with an initial roughness $R_a = 1.85–0.76$ mkm.

1.5 Development of a model of the quality of the appearance of paint and varnish coatings

To create a model that takes into account the influence of factors that determine the quality of the appearance of coatings, various methods of experiment planning are needed. In this case, as a rule, the form of the mathematical model is a complete form of the previously chosen regression equation, or its fragment. For example, a model of the form

$$y = a_1x_1 + a_2x_2 \tag{1.23}$$

or

$$y = a_{11}x_1^2 + a_{22}x_2^2 + a_{12}x_1x_2 \tag{1.24}$$

The exclusion of certain members from the model is associated with rigorous procedures for determining the significance of the coefficients.

Such an approach is universal for any output factor y and does not require any additional knowledge about the studied factor.

The following are some of the shortcomings of this approach:

- There is a possibility of output of the calculated value for physical and logical limits during interpolation of the model.
- The absurd results can be observed when substituting zero values separately for factors x_1 and x_2, even if these conditions were within a given interval of their experimental study.

It is of interest to consider, when developing a quality of model of appearance of coatings, the application of experimental planning methods based on combining probabilistic and deterministic approaches. In our opinion, the most optimal for describing the model is the multifactorial function of Protodyakonov of the following kind [18]:

$$Y = \frac{1}{y_{aver}} y_1(x_1) y_2(x_2) \tag{1.25}$$

where $Y_1(X_1)$, $Y_2(X_2)$ refers to particular algebraic dependencies on the factors X_1 and X_2, particular algebraic dependencies on the factors, and y_{aver} is the average value of all the experimental results considered.

Table 1.16 A two-factor experiment at three levels

N experiment	X_1	X_2
1	1	1
2	1	2
3	1	3
4	2	1
5	2	2
6	2	3
7	3	1
8	3	2
9	3	3

As a structural basis, the following experiment plan is proposed (Table 1.16).

Each line of the plan represents the specific conditions and results of the experiment. An important advantage in this case is that there are no restrictions on the kind of particular algebraic dependence.

Under conditions of uncertainty, it is possible to use additional a priori information related to the physical meaning of the studied dependence (increase, decrease, extremality, or certain boundary conditions).

If there is no physical concept regarding the influence of this factor or its nature is unclear, then it is necessary to limit the processing of the obtained data by the least squares method to the equation of the straight line. If necessary, the adequacy of particular dependencies is checked.

Another important advantage of the model (1.25) is that the value of the multifactor dependence corresponds exactly to the zero value of the partial dependence. The influence of particular functions on the multifactor dependence can be perceived as a shift of it from the mean value in proportion to the relative change in the particular functions.

Below are the results of the experiment and the calculation of the quality model of the appearance in the example of a polymer coating, which depend on the surface tension of the paint and the quality of the cement substrate. The quality of the cement substrate was estimated using the surface porosity index:

$$P = \frac{S_{por}}{S_{subst}} \times 100\% \qquad (1.26)$$

where S_{por} is the area of pores on the surface of the substrate, cm^2 and S_{subst} is the total surface area of the substrate, cm^2.

Variable factors were x_1, the surface porosity of the substrate, %, and x_2, the surface tension, mJ/m^2, and output parameter Y, the surface roughness, which was assessed using the Model 283 profilometer. Table 1.17 shows the experimental data of various factors affecting the quality of appearance of polymer coating. Specific point dependencies are presented in Figs. 1.19–1.20.

Table 1.17 Influence of various factors on the quality of appearance polymer coating

x_1	x_2	$Y_{calc.}/Y_{exp.}$
0	40	40/41.299
0	50.8	33.3/32.793
0	60.63	43.5/45.289
4.1	40	48/50.944
4.1	50.8	41.3/40.452
4.1	60.63	56.55/55.865
11.9	40	72/69.293
11.9	50.8	51.2/55.021
11.9	60.63	73.5/75.987

Note: $Y_{calc.}$ is the roughness values of the coating surface, obtained by calculation in accordance with the model (1.25), and $Y_{exp.}$ is the surface roughness values of the coating obtained experimentally.

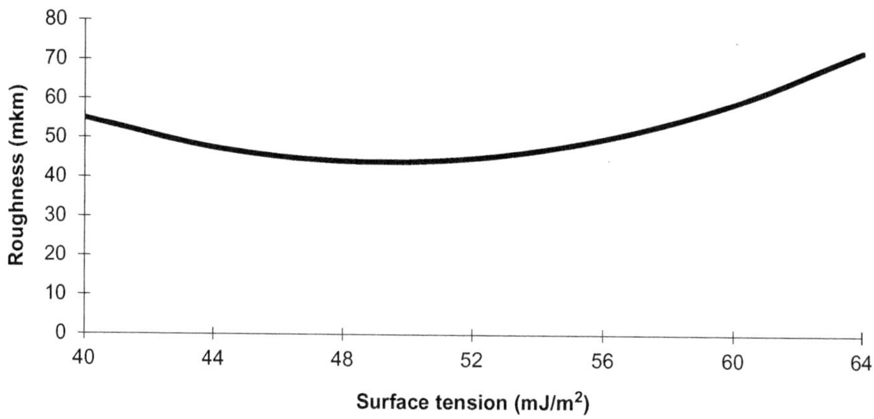

Figure 1.20 Influence of surface tension of polymer calcareous paint on the quality of the appearance of the coatings.

The results of the studies (Figs. 1.19–1.20) can be approximated by the following functions:

$$Y_2 = 359 - 12.83x_2 + 0.13x_2^2 \tag{1.27}$$

$$Y_1 = 39.15 + 2.23x_1 \tag{1.28}$$

The significance of particular dependencies was determined from the correlation coefficient. The Protodyakonov equation of significant private functions has the form

$$\frac{1}{51.039}(39,15 + 2,23x_1)(359 - 12,83x_2 + 0,13x_2^2) \tag{1.29}$$

and gives results that correlate well with the experimental data (Table 1.17). The correlation coefficient is 0.98. Thus, the Protodyakonov formula adequately describes the joint effect of the surface tension of the paint composition and the surface porosity of the substrate on the quality of the appearance of the coatings, taking into account their physical meaning.

1.6 Relationship of deformative properties of paint coatings with the roughness of their surface

Building and conserving the working condition of buildings and structures require a large number of paint and varnish compositions. Increasing competition in the market of finishing materials along with increasing demands of consumers require manufacturers to obtain high-quality painted surfaces. However, the practice of finishing works shows that often the quality of finishes is bad and leads to unscheduled repairs and additional costs [19]. In accordance with the statistical theory of the strength of solids, the probability of destruction of coatings is determined by the presence and concentration of defects, including on the surface of the coatings. Consequently, the quality of the appearance of coatings determines their stress state and resistance in the process of exploitation [19–21]. In Ref. [19] regularities of the formation of the quality of the appearance of coatings on a metallic substrate were established from the rheological properties of the paint and its ability to flow on surface of substrate. As noted in Ref. [21], the destruction of coatings on a metal substrate begins with different defects. Coatings on a porous cement substrate have their own peculiarities. The porous structure of the cement substrate affects the formation of the quality of the appearance of the coatings, which are characterized as lower quality. This, of course, has an effect on the resistance of coatings during operation. An analysis of the scientific and technical literature shows that the durability of coatings on a cement substrate has not been studied sufficiently.

In the study used the following paint: alkyd enamel PF-115 grade, oil paint brands MA-15, polystyrene paint brands PS-160, and acrylate paint universal. The surface quality of the paintwork was evaluated a roughness, which is determined by profilograph TR-100.

Assessment deformation of the coatings was carried out with the help of a tensile machine IR 5057-50 with the samples after 28 days of curing. With this method the sample is stretched until it ruptures (deformation speed of 1 mm/min). The $1 \times 1 \times 5$ cm samples were fixed in the clips of the tensile machine so that their longitudinal axis was in the direction of the stretching, and the force was applied equally all over the sample section. The tests were carried out at the temperature of 20°C and relative air humidity of 60%. The ultimate tensile strength estimation was carried out for four samples. The ultimate tensile strength R_{kog} for each sample was calculated by the following formula:

$$R_{kog} = \frac{F_{Pi}}{S_{Oi}} \quad (1.30)$$

where F_{Pi} is the stretching loading at the time of a rupture, N, and S_{Oi} is the initial cross-sectional area of a sample, mm^2.

The modulus of elasticity was calculated according to the chart "tension-deformation."

The modulus of elasticity for each sample (E_{upr}) in MPa was calculated by:

$$E_{upr} = \frac{R'_{kogi}}{\varepsilon'_i} \times 100 \quad (1.31)$$

where R'_{kogi} is the ultimate tensile strength at the time of the tangent separation from the chart "tension-deformation," MPa, and ε'_i is the relative lengthening at the time of rupture, %.

The presence of defects on the surface of the coatings will undoubtedly affect the physicomechanical properties of paint coatings. As can be seen, coatings have the elastoplastic character of the destruction (Tables 1.18 and 1.19).

Regardless of the type of paint, the strength and relative deformations are reduced, the plastic deformation is increased, and the elastic deformations are reduced with increasing roughness of surface (Figs. 1.21 and 1.22). Thus, when the surface of coatings roughness based on paint PS-160 is $R_a = 0.74$ mkm, the tensile strength R_p is 69.4 kgf/cm^2, the relative strain $\varepsilon_{rel} = 3.1\%$, with a roughness $R_a = 0.86$ mkm is $R_p = 50.5$ kgf/cm^2 and 1.75%. At a roughness of the film based on the paint PF-115 $R_a = 0.74$ mkm, the tensile strength R_p is 57.7 kgf/cm^2 and the relative strain $\varepsilon_{rel} = 44.3\%$, at a roughness $R_a = 1.74$ mkm is $R_p = 44.1$ kgf/cm^2 and 23%.

The dependence of the tensile strength on the roughness of the surface of the films can be approximated by an expression of the form:

$$R_p = a \times e^{b\,R_a} \quad (1.32)$$

where R_a is the surface roughness, mkm; b is a coefficient that takes into account the degree of reduction in strength from roughness, mkm^{-1}; and A is the coefficient that characterizes the value of tensile strength, at $R_a = 0$ (ideal model).

The quality of coatings is determined by the rheological properties of the paint and the porosity of the substrate. The presence of inclusions, shagreen, streaks, and waviness on the surface of coatings determines their stress state and endurance during operation. Figs. 1.23–1.24 shows the results of measuring the surface roughness of coatings during the moistening process. As can be seen, irrespective of the rheological properties of the paints and the porosity of the substrates during the moistening of the coatings in the first stage (up to 30 days), the roughness of the surface is reduced, that is, the surface microrelief is leveled due to the plasticizing effect of moisture (swelling of the coatings). In the future, due to the destructive effect of moisture, the surface roughness increases, caused by the appearance of microcracks, rashes, and bubbles.

Table 1.18 Deformations of films based on polystyrene paint PS-160 depending on surface roughness

Coating surface roughness, R_a (mkm)	**Strength at stretching, R_p (kgf/cm^2)**	**Relative deformation ε_{rel} (%)**	**Elastic deformation ε_{el} (%)**	**Plastic deformation ε_{pl} (%)**	**Share of elastic component deformation ε'_{el}**	**Share of plastic component deformation ε'_{pl}**
0.74	69.4	3.1	2.9	0.2	0.935	0.065
0.77	62.8	1.86	1.63	0.23	0.876	0.124
0.8	56.5	1.8	1.5	0.3	0.833	0.167
0.86	50.5	1.75	1.44	0.31	0.82	0.18
1.2	47.2	1.4	1.1	0.3	0.7857	0.2143

Table 1.19 Deformations of films based on paint PF-115 depending on surface roughness

Coating surface roughness, R_a (mkm)	**Strength at stretching, R_p (kgf/cm^2)**	**Relative deformation ε_{rel} (%)**	**Elastic deformation ε_{el} (%)**	**Plastic deformation ε_{pl} (%)**	**Share of elastic component deformation ε'_{el}**	**Share of plastic component deformation ε'_{pl}**
1.2	57.7	44.3	21.3	23	0.481	0.519
1.37	56.1	38	13.9	24.1	0.367	0.633
1.45	54.3	28	9	19	0.321	0.679
1.54	45.9	24	7	17	0.292	0.708
1.74	44.1	23	6.44	16.56	0.281	0.719

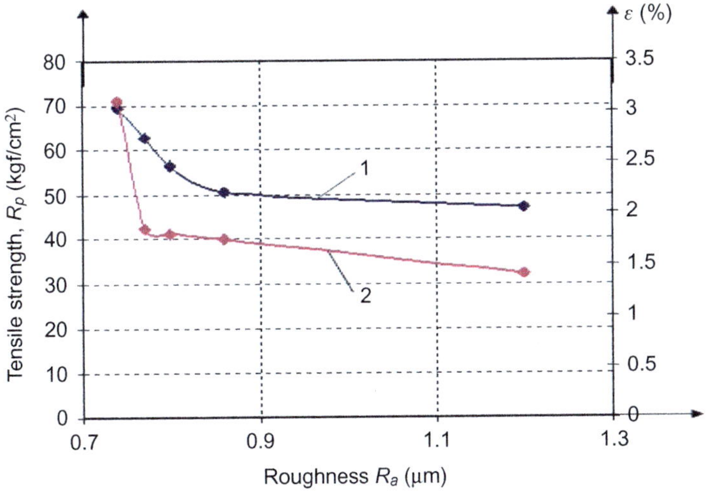

Figure 1.21 Dependence of the tensile strength (1) and the relative elongation (2) on the roughness of the film surface on the basis of paint PS-160.

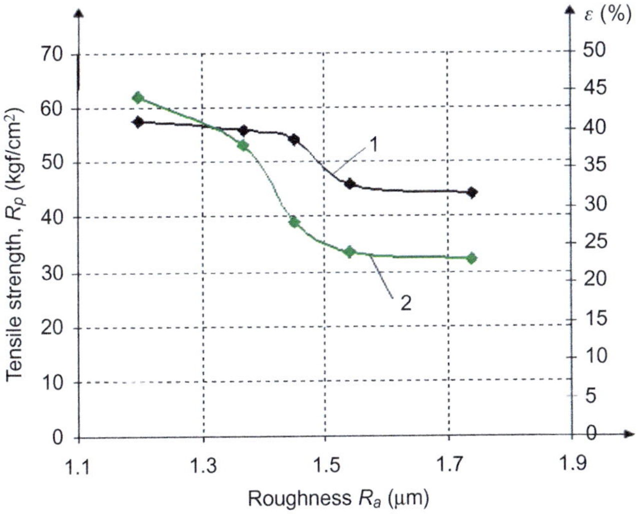

Figure 1.22 Dependence of the tensile strength (1) and the relative elongation (2) on the roughness of the film surface on the basis of paint PF-115.

A more stressed state of the coating in places of greater roughness contribute to destroy the coating in these places during operation. In the process of cyclic freezing-thawing the cracks appear locally and are formed near defects on the surface of the coating. In particular, on a coating based on paint MA-15 with a roughness $R_a = 0.23$ mkm cracks appeared after 5 freeze-thaw cycles, and on a coating with a roughness $R_a = 0.14$ mkm, after 15 cycles tests.

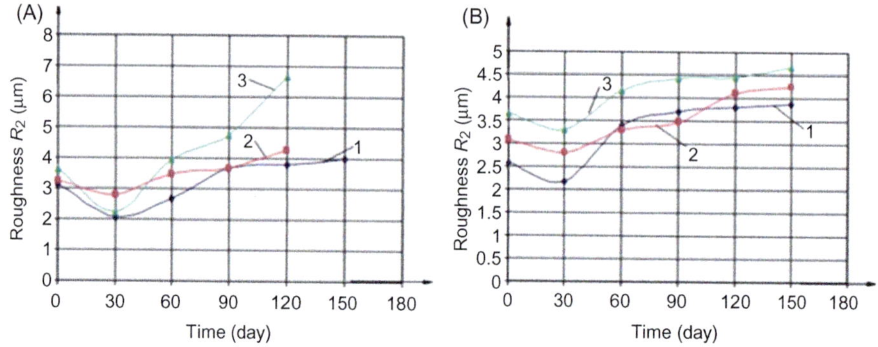

Figure 1.23 Change of the roughness of coatings in the process of moistening.
(A) Based on paint PF-115 (a method of applying, brush, porosity 24%): 1—viscosity is 0.001×10^3 Pa s; 2—viscosity is 0.00065×10^3 Pa s; 3—viscosity is 0.00026×10^3 Pa s
(B) Based on dispersion paint (a method of applying brush, on the surface of the putty): 1—viscosity is 0.0347×10^3 Pa s; 2—viscosity is 0.02317×10^3 Pa s; 3—viscosity is 0.013×10^3 Pa s.

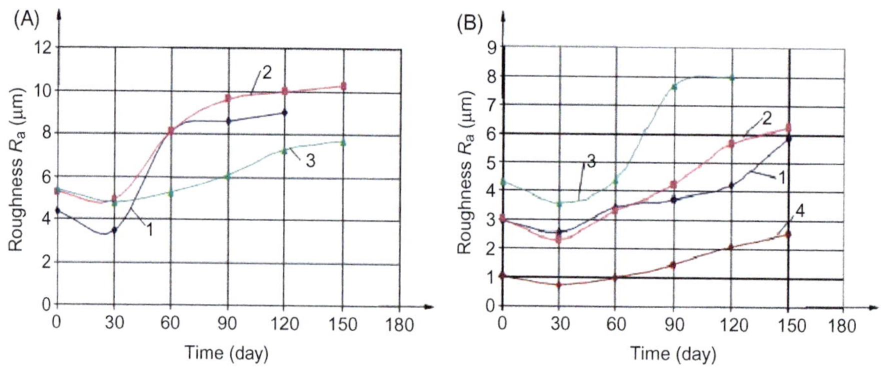

Figure 1.24 Change of the roughness of coatings in the process of moistening.
(A) Based on paint MA-15 (airless method, porosity 24%): 1—viscosity is $\eta_1 = 0.0026 \times 10^3$ Pa s; 2—viscosity is $\eta_2 = 0.0020 \times 10^3$ Pa s; 3—viscosity is $\eta_3 = 0.0014 \times 10^3$ Pa s (B) Based on paint PF-115 (method pouring): 1—viscosity is $0.001.10^3$ Pa s, porosity 24%; 2—viscosity is 0.001×10^3 Pa s, porosity 28%; 3—viscosity is 0.001×10^3 Pa s, porosity 32%; 4—viscosity is 0.001×10^3 Pa s, on the surface of the putty.

Similar regularities are also seen for other coatings. The obtained data correlate well with other indices of the protective and decorative properties of the coating (color change, gloss variation, chalking, mud retention, bronzing, etc.).

The elastoplastic characteristic of the destruction of the coatings studied was also found. The models of the strength of coatings depending on the roughness of surface are given. As can be seen, irrespective of the rheological properties of paint and the porosity of the substrates during the moistening of the coatings in the first stage, the roughness of the surface is reduced and the surface microrelief is leveled due to the plasticizing effect of moisture (swelling of the coatings). In the future, due to the destructive effect of moisture, the surface roughness will increase. A more stressed state of the coating in places of greater roughness contribute to destroy the coating in these places during exploitation [22,23]. In the process of cyclic freezing-thawing the cracks appear locally and are formed near defects on the surface of the coating.

The obtained data correlate well with other indices of the protective and decorative properties of the coating (color change, gloss variation, chalking, mud retention, bronzing, etc.).

This research can be used to develop recommendations for increasing the resistance of coatings and to select the optimum rheological properties of paints depending on the porosity of the substrate.

1.7 Accounting for the hereditary factor when assessing the quality change of the appearance of paint and varnish in the aging process

In accordance with the statistical theory, the probability of destruction of protective and decorative coatings depends on the presence of defects on their surface, which can be determined by the relation [22]

$$P = 1 - \exp(-\rho s) \tag{1.33}$$

where ρ is the concentration of defects and s is the surface area.

Obviously, with the same concentration of defects, the probability of destruction of coatings increases with increasing surface area. Therefore, coatings with a rough surface will be destroyed at a higher rate.

During the exploitation of the coating, when the values of the parameters ρ and s in formula (1.33) increase, the probability of destruction increases.

The intensity of destruction under the influence of climatic factors is not the same at different stages of operation. This is mainly due to changes in the structure of the coating. It was established [24] that during a certain incubation period (the duration of which depends on the type of coating), there is an insignificant increase in defectiveness. Then, in the active stage of damage accumulation, significant changes occur in the structure and properties of the coating (color change, shine, cracking, etc.).

In many cases, the kinetics of damage accumulation is described by a first-order differential equation:

$$\frac{dW}{dt} = K(1 - W) \tag{1.34}$$

where k is the rate constant of damage accumulation.

A common solution is the function

$$W = 1 - Ce^{-kt} \tag{1.35}$$

If at the time of the start of operation there are already damages due to structural defects in the coating, that is, $W(0) = W_0$, then the desired dependence is a solution of the Cauchy problem in the form

$$W = 1 - (1 - W_0)e^{-kt} \tag{1.36}$$

It is known [24] that in the incubation period and in the active stage of damage accumulation, the rate of damage accumulation is different. Thus, for PVAC coating it is 0.00163 and 0.016 cycles^{-1}, respectively. For KO-168 coatings, the constant k in the incubation period is 0.00163 cycles^{-1} and in the active stage 0.011 cycles^{-1}.

Similar conclusions can be drawn by the change in the aging process of the surface area of the coating. As is known, under the influence of climatic factors, the surface layers of the coating are destroyed, leading to an increase in the roughness, and consequently, the surface area of the coating. In addition, the increase in surface area is associated with the resulting macro- and microcracks.

Consider moisturizing as one of the particular aging factors. Analysis of the experimental data indicates that the beginning of the active stage of accumulation of lesions coincides in time with the first significant changes in the surface area of the coating (Figs. 1.25–1.26).

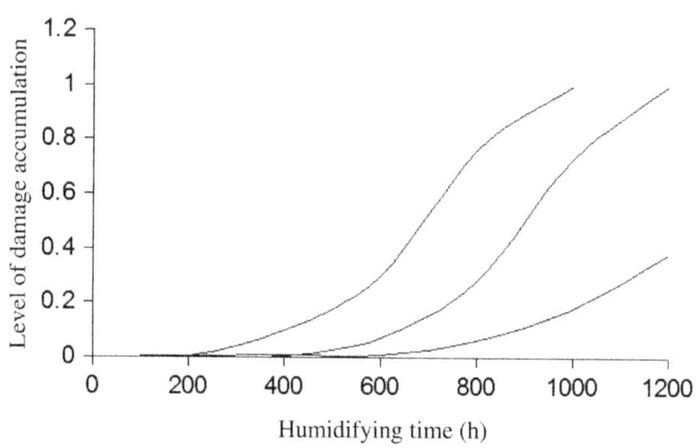

Figure 1.25 Change in the level of accumulation of damage to coatings during aging:
1—PVAC
2—polymer-calcareous coating
3—lime coating.

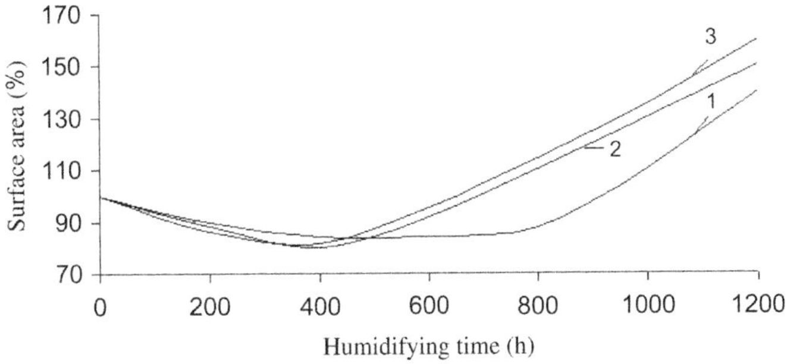

Figure 1.26 Change in surface area of coatings when moistened:
1—PVAC
2—polymer-calcareous coating
3—lime coating.

Thus, the kinetics of the destruction of coatings from a mathematical point of view should be described by differential equations of order higher than the first. This means that when analyzing a process, it is necessary to take into account not only the rate of change of the monitored parameter, but also at least the acceleration. Therefore, it is proposed to use, as a model of the fracture process,

$$Y(t) = C_1 e^{\alpha_1 t} + C_2 e^{\alpha_2 t} + C_3 e^{\alpha_3 t} \tag{1.37}$$

which is a solution of a third-order differential equation.

The coefficients α_1, α_2, α_3 characterize the rate of change of the controlled parameter at different stages of coating aging.

Analyzing the empirical curves for the kinetics of damage accumulation, we can assume that $\alpha_1 < < \alpha_2$ (for the surface area of the coating, it is even characteristic $\alpha_1 < 0$), $\alpha_2 > \alpha_3$.

As an example, consider the change in the surface area of a PVAC during the moisture process. From the experimental data follows $y(0) = 100$, $y(800) = 95$, $y(1200) = 140$, and $y'(800) = 0$.

Taking into account (1.37), we identify the parameters of the model:

$$\begin{aligned}
&C_1 + C_2 + C_3 = 100 \\
&C_1 e^{800\alpha_1} + C_2 e^{800\alpha_2} + C_3 e^{800\alpha_3} = 95 \\
&C_1 e^{1200\alpha_1} + C_2 e^{1200\alpha_2} + C_3 e^{1200\alpha_3} = 140 \\
&\alpha_1 C_1 e^{800\alpha_1} + \alpha_2 C_2 e^{800\alpha_2} + \alpha_3 C_3 e^{800\alpha_3} = 0
\end{aligned} \tag{1.38}$$

From the first three equations we obtain

$$C_1 = \frac{100e^{800\alpha_2+1200\alpha_3} - 100e^{800\alpha_3+1200\alpha_2} - 95e^{1200\alpha_3} + 140e^{800\alpha_3} + 95e^{1200\alpha_2} - 140e^{800\alpha_2}}{e^{800\alpha_2+1200\alpha_3} - e^{800\alpha_3+1200\alpha_2} - e^{800\alpha_1+1200\alpha_3} + e^{800\alpha_3+1200\alpha_1} + e^{800\alpha_1+1200\alpha_2} - e^{800\alpha_2+1200\alpha_1}}$$

$$C_2 = \frac{95e^{1200\alpha_3} - 140e^{800\alpha_3} - 100e^{800\alpha_1+1200\alpha_3} + 100e^{800\alpha_3+1200\alpha_1} \pm 140e^{800\alpha_1} - 95e^{1200\alpha_1}}{e^{800\alpha_2+1200\alpha_3} - e^{800\alpha_3+1200\alpha_2} - e^{800\alpha_1+1200\alpha_3} + e^{800\alpha_3+1200\alpha_1} + e^{800\alpha_1+1200\alpha_2} - e^{800\alpha_2+1200\alpha_1}}$$

$$C_3 = \frac{95e^{1200\alpha_3} - 140e^{800\alpha_3} - 100e^{800\alpha_1+1200\alpha_3} + 100e^{800\alpha_3+1200\alpha_1} \pm 140e^{800\alpha_1} - 95e^{1200\alpha_1}}{e^{800\alpha_2+1200\alpha_3} - e^{800\alpha_3+1200\alpha_2} - e^{800\alpha_1+1200\alpha_3} + e^{800\alpha_3+1200\alpha_1} + e^{800\alpha_1+1200\alpha_2} - e^{800\alpha_2+1200\alpha_1}}$$

(1.39)

We also determine α_1 and α_3 from the experimental characteristic $Y\,(t)$ (Fig. 1.30). Since $\alpha_1 < < \alpha_2$, the component $\lambda_2(x_m/\lambda - \lambda_{21})C_1e^{-\lambda_{1}t}$ determines the process until the end of the incubation period, that is, at $0 \le t \le 800$. The value can be determined at the end of the experimentally obtained process $Y\,(t)$ [ ].

$$Y(0+T) = Ae^{\alpha_1(0+T)}, \text{ При } t < 800 \quad (1.40)$$

$$y(t+T) = Ae^{-\lambda_2(t+T)}, \text{ При } t > > 800 \quad (1.41)$$

$$\alpha_1 = \frac{\ln(Y(T)/Y(0))}{T} \quad (1.42)$$

$$\alpha_3 = \frac{\ln(Y(t+T)/Y(t))}{T} \quad (1.43)$$

In this case, $\alpha_1 = -0.000064$, $\alpha_3 = 0.0012$.

From the fourth equation of system (1.39) we obtain an implicit expression for $<\alpha_2$:

$$-0.00006C_1 + \alpha_2 C_2 e^{800\alpha_2} + 0.003C_3 = 0 \quad (1.44)$$

An approximate solution of this equation gives $\alpha_2 = 0.011$.

As a result, the dependence (1.38) is represented in the form

$$Y(t) = 100.364e^{-0.000064t} + 0.00009e^{0.011t} - 0.364e^{0.0012t} \quad (1.45)$$

From the above, the following algorithm for identifying kinetic processes of a given type follows:

1. According to the initial changes in the incubation period α_1, it is determined, at the end of the empirical curve—α_3;
2. Constants C_1, C_2, C_3 are represented as functions;
3. On the characteristic point of the empirical curve, Eq. (1.37) is constructed, an approximate value of the constant is found;

4. are identify

$$Y(t) = C_1 e^{\alpha_1 t} + C_2 e^{\alpha_2 t} + C_3 e^{\alpha_3 t} \tag{1.46}$$

The proposed algorithm can be used to solve other problems of building material science with the possibility of describing the processes under consideration as solutions of differential equations of the third order.

Considering on the basis of this function the hereditary component

$$I(t) = \int_0^t K(t, \tau) \frac{dY(\tau)}{d\tau} \tag{1.47}$$

we can conclude, that in the stages of aging of the coating described above, it is constructed additively for each period separately. In the incubation period, its influence is positive, and in the active stage it aggravates the aging process. Carrying out calculations for other coatings, we have the data shown in Table 1.20.

Applying the principles of the hereditary theory of aging to these dependences, it is possible to calculate the influence of the hereditary factor on the change in the surface area of the coatings under moistening. Indeed, when the coatings are moistening in the first stage (from 300 to 800 hours of moistening, depending on the coating), the surface microrelief is leveled due to the plasticizing effect of moisture. Subsequently, an active stage of accumulation of lesions occurs along with an increase in the surface area.

The calculated data allow us to assume that at the first stage the influence of the hereditary factor (i.e., the duration of the plasticizing effect of moisture) on the change in the surface area of the coating is codirected with the action of other aging factors, and the share of this influence in the overall change in the controlled parameter is very significant.

So, for polymer lime coating it's 25%, for lime increasing to 42%. At the beginning of the active stage of accumulation of damages (600–800 hours of moistening) the hereditary factor aggravates the aging process. The destructive effect of moisture at a given time has a basic effect on the destruction of the coating. Subsequently, the microcracks of the coating increase, the internal stresses in the material fall off, and the accumulation of damages occurs at a significantly lower rate. Destructuring leads to the fact that the accumulated damages practically do not affect each other, and the hereditary properties of damage are extremely weak (Tables 1.21–1.26).

Table 1.20 Calculated dependencies of the area change in surface coatings when moistened

Lime coating	$Y(t) = 129,292e^{-0,00017t} + 14,06e^{0,002t} - 43,325e^{0,0007t}$
Polymer-calcareous coating	$Y(t) = 110,059e^{-0,00021t} + 2,089e^{0,003t} - 12,898e^{0,0091t}$
Coating PVAC	$Y(t) = 100,364e^{-0,000064t} + 0,00009e^{0,011t} - 0,364e^{0,0012t}$

Table 1.21 Change in the surface area of the coating when moistened

Lime coating	t	The calculated value of the surface area Y	Primary importance Y^*	Hereditary component $I_{нф}$	I—incubation period	II—the first period of the active stage	III—the second period of the active stage
					I Y^*	II Y^*	III Y^*
					I $I_{нф}$	II $I_{нф}$	III $I_{нф}$
	100	97,79	98,042	−0,105	126,691	17,137	−45,786
					−0,125	0,135	−0,115
	200	96,078	96,674	−0,401	124,143	20,887	−48,356
					−0,479	0,521	−0,444
	300	95	96,033	−0,86	121,646	25,458	−51,071
					−1,029	1,136	−0,967
	400	94,723	96,29	−1,457	119,199	31,029	−53,938
					−1,748	1,957	−1,665
	600	97,455	100,384	−2,98	114,452	46,096	−60,164
					−3,596	4,141	−3,524
	800	106,596	111,265	−4,822	109,894	68,479	−67,109
					−5,855	6,938	−5,904
	1000	125,669	132,393	−6,867	105,517	101,731	−74,855
					−8,392	10,236	−8,711
	1200	159,999	168,948	−9,026	103,395	151,129	−83,496
					−11,104	13,946	−11,868

Table1.22 Change in the surface area of the coating when moistened

Polymer-calcareous coating	t	The calculated value of the surface area Y	Primary importance Y^*	Hereditary component $I_{нф}$	I—incubation period	II—the first period of the active stage	III—the second period of the active stage
					I Y^*	II Y^*	III Y^*
					I $I_{нф}$	II $I_{нф}$	III $I_{нф}$
	100	97,47	97,892	−0,106	108,121	3,815	−14,043
					−0,093	0,042	−0,054
	200	95,214	96,071	−0,407	106,216	5,181	−15,326
					−0,358	0,161	−0,211
	300	93,341	94,656	−0,878	104,345	7,034	−16,726
					−0,77	0,351	−0,459
	400	92	93,811	−1,497	102,508	9,557	−18,253
					−1,31	0,605	−0,791
	600	91,809	94,816	−3,099	98,928	17,627	−21,74
					−2,705	1,28	−1,674
	800	97,354	102,095	−5,08	95,474	32,513	−25,892
					−4,419	2,144	−2,804
	1000	113,664	121,272	−7,33	92,141	59,969	−30,838
					−6,356	3,163	−4,137
	1200	149,988	162,807	−9,765	88,924	110,612	−36,728
					−8,438	4,309	−5,637

Table 1.23 Change in the surface area of the coating when moistened

PVAC coating	t	The calculated value of the surface area Y	Primary importance Y^*	Hereditary component $I_{нф}$	I—incubation period	II—the first period of the active stage	III—the second period of the active stage
					I Y^*	II Y^*	III Y^*
					I $I_{нф}$	II $I_{нф}$	III $I_{нф}$
	100	99,314	99,445	−0,027	99,828	$2{,}728 \times 10^{-4}$	−0,381
					−0,026	$4{,}828 \times 10^{-6}$	$-7{,}921 \times 10^{-4}$
	200	98,626	98,893	−0,103	99,29	$8{,}279 \times 10^{-4}$	−0,398
					−0,1	$1{,}87 \times 10^{-5}$	$-3{,}067 \times 10^{-3}$
	300	97,936	98,343	−0,225	98,757	$2{,}512 \times 10^{-3}$	−0,417
					−0,278	$4{,}074 \times 10^{-5}$	$-6{,}684 \times 10^{-3}$
	400	97,246	97,799	−0,386	98,227	$7{,}623 \times 10^{-3}$	−0,436
					−0,374	$7{,}016 \times 10^{-5}$	−0,012
	600	95,901	96,769	−0,81	97,176	0,07	−0,477
					−0,786	$1{,}485 \times 10^{-4}$	−0,024
	800	95	96,261	−1,347	6,136	0,646	−0,522
					−1,306	$2{,}489 \times 10^{-4}$	−0,041
	1000	98,314	100,487	−1,972	95,107	5,95	−0,571
					−1,912	$3{,}672 \times 10^{-4}$	−0,06
	1200	139,998	148,25	−2,665	94,089	54,785	−0,625
					−2,583	$5{,}002 \times 10^{-4}$	−0,082

Table 1.24 Share of change (%) due to the hereditary factor in the total change in the surface area of the lime coating during the moistened process

Lime coating	Time (h)	I	II	III
	100	−0,099		
	200	−0,386		
	300	−0,846		
	400	−1,467	31,562	
	600		44,961	
	800		50,708	29,638
	1000			39,205
	1200			47,885
	The share of change due to the hereditary factor in the total change for phase	−42%	+28%	+7,8%

Table1.25 Share of change (%) due to the hereditary factor in the total change in the surface area of the polymer- calcareous coating during the moistened process

Polymer-calcareous coating	Time (h)	I	II	III
	100	−0,086		
	200	−0,337		
	300	−0,738		
	400	−1,278	6,327	
	600		7,259	
	800		6,594	10,83
	1000			13,417
	1200			15,347
	The share of change due to the hereditary factor in the total change for phase	−25,4%	+66,9%	+8,9%

Table 1.26 Share of change (%) due to the hereditary factor in the total change in the surface area of the polymer-calcareous coating during the moistened process

PVAC coating	Time (h)	I	II	III
	100	−0,026		
	200	−0,101		
	300	−0,221		
	400	−0,381		
	600	−0,809		
	800	−1,359	0,039	
	1000		$8,171 \times 10^{-3}$	10,554
	1200			13,141
	The share of change due to the hereditary factor in the total change for phase	−30,6%	+48,1%	+2,9%

References

[1] L.P. Orentlicher, V.I. Loganina, Protective Decorative Coatings for Concrete and Stone Buildings. Study guide, Stroyizdat, Moscow, 1992.
[2] V.I. Loganina, L.V. Makarova, R.V. Tarasov, Method of assessment quality protective and decorative coating concrete cement. Case Stud. Constr. Mater. 4 (2016), 81–84. Available from: https://doi.org/10.1016/j.cscm.2016.01.003.
[3] Application of Fractals and Chaos, Springer-Verlag, Berlin, 1993.
[4] E. Feder, Fractals. Trans. with the English-M.: Mir, 254s, Jens Feder, Plenum Press, New York, 1991. 1988.
[5] B.C. Ivanova, I.R. Kuzeyev, M. Sikirica, Synergetics and fractals, Universality of Mechanical Behavi B. B. or of Materials, Ufa, publishing house of USPTU, 1998. 273 p.
[6] B.B. Mandelbrot, The Fractal Ceometiy of Natire Текст./В. B. Mandelbrot, W.H. Freeman and Company, New York, 1988. 107 p.
[7] A.A. Bykhovsky, Spreading, Kiev, Nauk. Dumka, 1983, 191 p.
[8] V.I. Loganina, L.V. Makarova, Technique of the assessment of crack resistance of the protective decorative coatings, Contemp. Eng. Sci. 7 (36) (2014) 1967–1973. Available from: https://doi.org/10.12988/ces.2014.411239.
[9] L.A. Sukhareva, The Durability of Coatings, Moscow, Chemistry, 1984. 240 p.
[10] V.I. Loganina, Maintenance of quality of paint and varnish coverings of building products and designs, Contemp. Eng. Sci. 7 (36) (2014) 1943–1947. HIKARI Ltd, www.m-hikari.com http://dx.doi.org/10.12988/ces.2014.411243.
[11] E. Shindovsky, O. Shyurts, Statisticheskie management techniques kachestvom. M: Peace, 1976.
[12] The GOST R 50779.30 - 95 Statistical Methods. Acceptance Quality Control. General Requirements, Moscow, Publishing House of Standards, 1995.
[13] GOST R 50779.50 - 95 Statistical Methods. Acceptance Quality Control by Variables. General Requirements. Moscow, Publishing House of Standards, 1995.
[14] M.I. Karyakina, Testing of Paint and Varnish Materials and Coatings, Chemistry, Moscow, 1988. 272 p.
[15] M.I. Karyakina, Physico-chemical Basis of the Formation and Aging of Coatings, 1980, Chemistry, Moscow, 1980. 216 p.
[16] E.A. Andryushchenko, Light Resistance of Paints, Chemistry, Moscow, 1986, p. 187.
[17] A.D. Zimon, Adhesion of Films and Coatings, Chemistry, Moscow, 1977. 351 p.
[18] M.M. Protodyakonov, R.I. Teder, Method of Rational Planning of the Experiment, Science, Moscow, 1970.
[19] V.I. Loganina, T.V. Uchaeva, P.V. Monastyrev, The method to estimate the surface appearance quality of the paint applied to the cement, J. Eng. Appl. Sci. 11 (11) (2016) 2409–2410.
[20] V.I. Loganina, Economic estimation of quality process of coloring building products and designs, Contemp. Eng. Sci. 8 (2) (2015) 71–75. HIKARI Ltd, www.m-hikari.com http://dx.doi.org/10.12988/ces.2015.412248.
[21] V.I. Loganina, Evaluation reliability of the control building materials and products for stability technological processes of production, Contemp. Eng. Sci. 7 (36) (2014) 1927–1933. HIKARI Ltd, www.m-hikari.com. Available from: https://doi.org/10.12988/ces.2014.411207.

[22] G.M. Bartenev, Yu.S. Zuev, Strength and Destruction of Highly Elastic Materials, Chemistry, Moscow-Leningrad, 1984.
[23] V.I. Loganina, The influence of surface quality of coatings on their deformation properties, Contemp. Eng. Sci. 7 (36) (2014) 1935–1941. HIKARI Ltd, www.m-hikari.com http://dx.doi.org/10.12988/ces.2014.411241.
[24] G.M. Bartenev, Strength and Fracture Mechanism of Polymers. Main. Chemistry, Chemistry, Moscow, 1984. 280 p.

Estimation of the stressed-deformed state of coating from the quality of their appearance

2

2.1 Method for assessing the stress state of coatings

Currently available methods of experimental study of the stress–strain state (SSS) of structures rely on direct measurement of deformations in the test article. Stresses are determined indirectly through deformations. Among the optical methods of analyzing the work of structures, a special place is occupied by new holographic methods. Holographic methods allow one to study the SSS of all kinds of natural materials and structures [1–6].

The essence of holographic nondestructive testing is that the position of the defect localization is determined by the anomaly of the pattern of interference fringes. Internal defects of the object are revealed from the pattern of the bands, if they cause sufficient perturbation of the field of deformations and displacements, and hence of the interference pattern on the observed surface of the sample.

Interferograms are typically obtained by comparing two states of the sample surface before and after the impact or loading. The internal paint defects are detected by the pattern of fringes if they cause enough disturbance of the displacement field, and thus of the interference pattern observed on the sample surface. The optical scheme of the hologram recording is shown in Fig. 2.1.

The objects of the research here were two painted samples of cement-sand matrix $4 \times 4 \times 16$ sm in size with different void content. The method used to obtain the holograms was as follows. The samples were placed in a box. In the vicinity of the coating surface a high-resolution plate PFG-03M sensitized to the red spectrum of the laser was rigidly fixed. To register holograms in colliding beams at a certain distance from the samples, a semiconductor laser on the red line (the laser wavelength $\lambda = 0.65$ mk) was used. The laser has sufficient coherence for recording holograms.

At the initial stage we made the first exposure, and then we added water to the box to 2/3 of the height of the mortar sample. After a certain period of time after the sample had been saturated with moisture, the second exposure was made, and then photoplates were photoprocessed with GP-3 developer, recommended by the manufacturer of the plates (without fixing) and then it was washed with distilled water. After the hologram was photoprocessed and dried, we studied the changes of the coating and its SSS [1].

Increasing the Durability of Paint and Varnish Coatings in Building Products and Construction.
DOI: https://doi.org/10.1016/B978-0-12-817046-5.00002-4

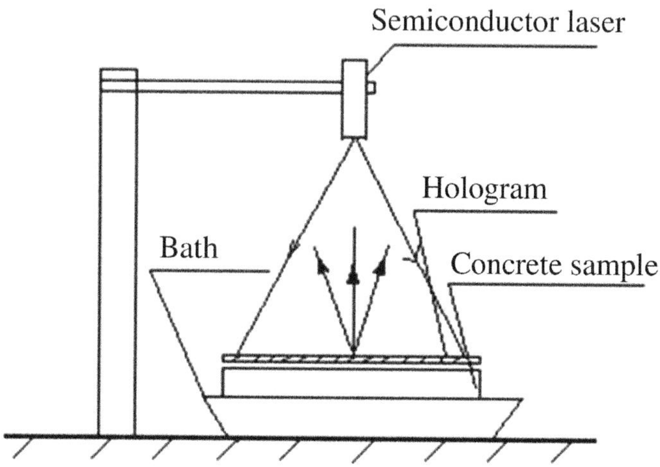

Figure 2.1 Optical scheme of registration of holograms.

The time between the first and second exposures may vary depending on the frequency of the interference fringes visible on the hologram. The higher the frequency, the smaller should be the time between exposures. It is important to change the hologram sensitivity for extracting necessary deformation components. In addition, a double-exposure hologram is convenient to filter, producing an optical differentiation in the optical filtering scheme of separate flat components.

2.2 Connection of the stress–strain state of paint coatings with the quality of their appearance

To establish the connection between the coating appearance and its state of stress we had to do the following experiment. Different paints were applied with a brush on substrate in two layers with a 24-hour in-between drying. The following paints were applied: alkyd enamel PF-115 grade, oil paint AI-15, acrylic paint "universal" and acrylic latex (exterior) paint. The difference in coating appearance quality was achieved by changing the porosity and rheological properties and was determined by the surface roughness R_a. The coating surface roughness was determined by profilograph TR-100 state. In total we performed 50 measurements on each surface on Scheme (Fig. 2.2).

Fig. 2.3 shows the photograph of the interferogram of the paint coating based on oil paint MA-15 with viscosity 0.0026×10^3 Pa s on a cement substrate with porosity of 24% (Fig. 2.3) and 32% (Fig. 2.4).

When analyzing the interferogram we can see that the greatest distortion of the interference fringe pattern is observed on the coating on the cement substrate with

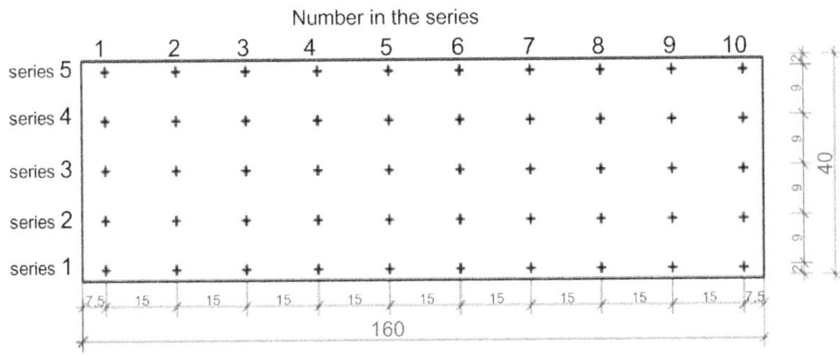

Figure 2.2 Scheme of the measurement of surface roughness paint coatings on cement substrate (all dimensions in mm).

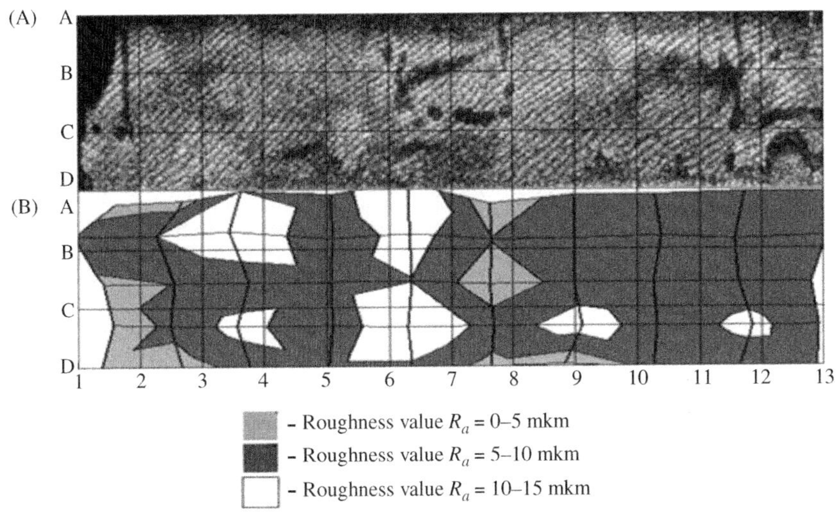

Figure 2.3 Interferogram of the paint coating based on paint MA-15: the porosity of the substrate 24% (A) and the distribution of roughness over the surface of the coating (B).

porosity $P = 32\%$ and surface condition $R_a = 7.76$ mkm. The most stress is observed on the coating with inclusions, waviness, drips and shagreen [7–9]. At these points there is a failure of the interference pattern, indicating the places of possible cracks. The same is true for other paints with different rheological properties.

Analysis of the experimental data indicates that there is a correlation between the roughness index of the coating surface and its SSS. For example, at roughness of the paint coating on the basis of MA-15 paint on a substrate with a porosity of 24%, equal to $R_a = 4–6$ mkm (Fig. 2.3) a disturbance of the order and the curvature of the bands on the interferogram in the area of the paint coating B3-A3-A4-B4 is observed. This indicates the presence in this area of possible

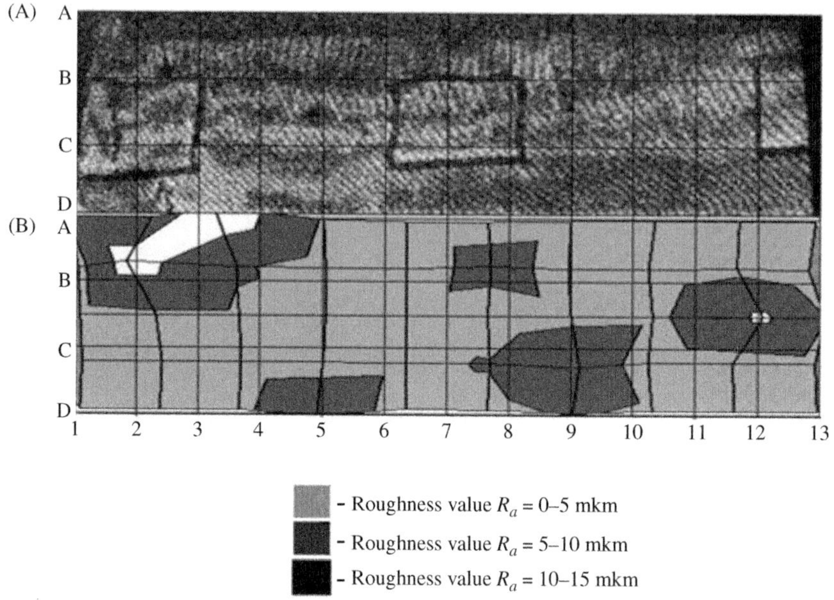

Figure 2.4 Interferogram of the paint coating based on paint MA-15: the porosity of the substrate 32% (A) and the distribution of roughness over the surface of the coating (B).

defective internal fracture and local stress concentration (2.3–2.5). In the D3-C3-C4-D4 and C4-D4-C5-D5 regions, the roughness of the coating is 8–12 mkm and also a disturbance in the order of the bands is observed in this region. Similar disturbances of the interference pattern are also observed in the regions D9-C9-C10-D10 and D11-C12-C12-D11, in which the roughness of the coating is 10–15 mkm. These regions were separately isolated and are shown in Fig. 2.5.

In the region of D3-C3-C4-D4 (Fig. 2.4), which is characterized by a low roughness value (0–2 mkm), a more even distribution of the interference fringes is observed. This indicates a uniform SSS of the coating in this zone. The region D3-C3-C4-D4 was separated separately and is shown in Fig. 2.6.

In the regions of D8-C8-C9-D9 and D9-C9-C10-D10 (Fig. 2.4), the roughness is 5–10 μm. There is a distortion of the pattern of bands and a violation of the order of the bands is observed. A similar pattern is also seen for the C8-B8-B9-C9 region. These regions were also isolated from Fig. 2.4 and are presented in Fig. 2.7.

We established the concentration of defects on the area of the coatings (the number of defective areas)-n_{def}. The coating is considered "nonfunctional" if on its surface there are more than n_{def} sections, determined from the expression

$$n_{def} = \frac{S_{def} n}{100} \tag{2.1}$$

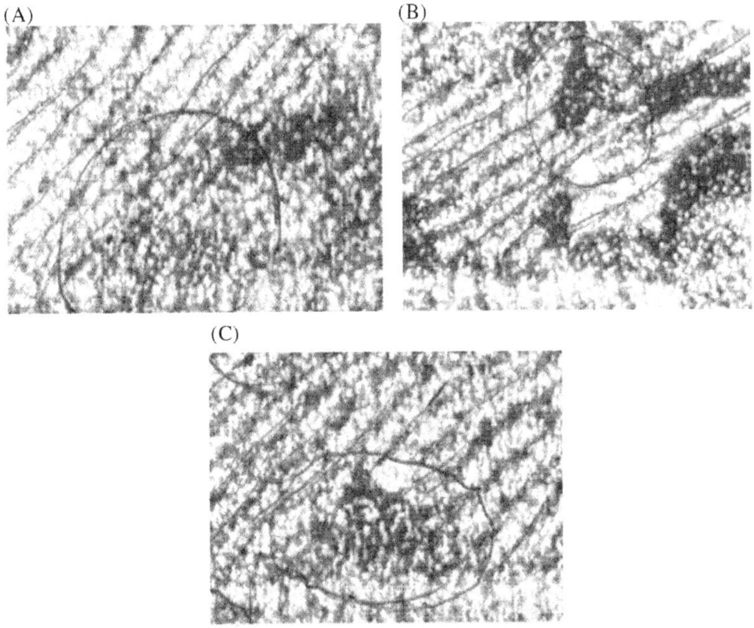

Figure 2.5 Zones of order violation and distortion of the interference pattern of the bands: (A) D3-C3-C4-D4 and C4-D4-C5-D5, (B) D11-C12-C12-D11 and (C) D9-C9-C10-D10 (MA-15 with a dynamic viscosity of 26.0 Pa s and a substrate porosity of 24%).

In accordance with law, the probability in the Poisson distribution of failure-free operation we established:

$$F(c) = \sum_{x=0}^{S_{def}n} \frac{c^x}{x!} e^{-c} \tag{2.2}$$

where $c = np = \text{const}$.

The results of the studies are given in Tables 2.1 and 2.2.

A more tense state of the coating in places of greater roughness promotes to destroy the coating in these places during operation. As is known, in the process of cyclic freezing-thawing the cracks appear locally and are formed near defects on the surface of the coating. In particular, surface cracks visible to the naked eye appeared on the paint coating MA-15, which were characterized by a roughness index $R_a > 3$ mkm, after 10 cycles of freeze-thaw and on a coating with a roughness index $R_a < 3$ mkm after 15 test cycles.

Similar patterns are obtained when tested in alternate moistening-drying. The dynamics of fracture coatings in the process of wetting and drying are shown in Table 2.2.

The results (Tables 2.1 and 2.2) suggest that there is a proportional relationship between the original roughness of the coating and the likelihood of its destruction.

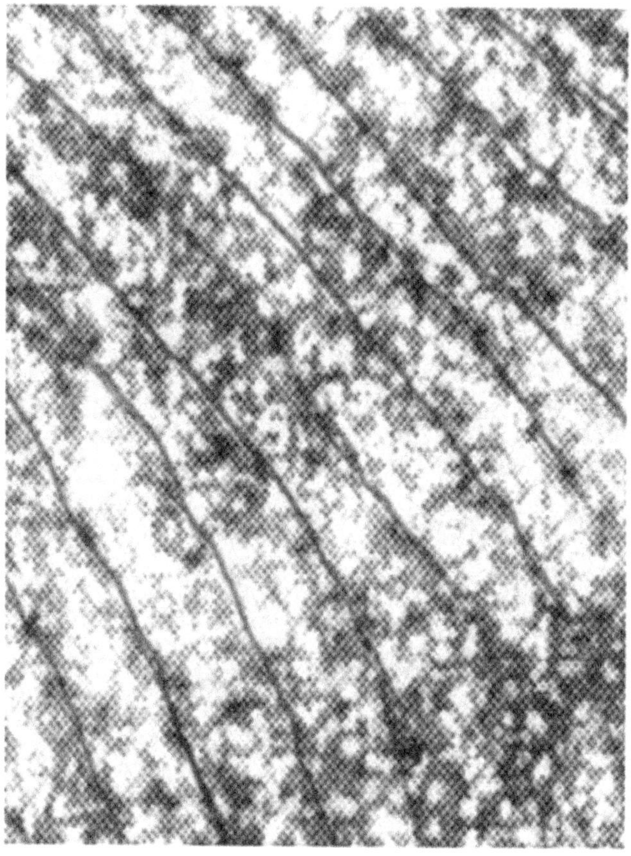

Figure 2.6 Zone of uniform distribution of the bands D3-C3-C4-D4 (paint MA-15, dynamic viscosity 26.0 Pa s and porosity of the substrate 32%).

Thus, the probability of failure of the coating based on latex paint with an initial roughness $R_a = 2.55$ mkm after 5 test cycles, 34.3%, and coverage with the original roughness $R_a = 3.29$ mkm is 36.8%.

Fig. 2.8 shows an interferogram of the SSS of coatings based on paints MA-15, applied by brush on substrate with porosity 28% and with a dynamic viscosity of 26.0 Pa s after 60 days of moistening. On the interferogram (Fig. 2.8), the complex spatial displacements of the crack opening and the high concentration of stresses near the crack are traced, leading to the formation of a zone of initiation of plastic deformations [10–12].

The initial roughness of the coating was (to moisturize) $R_a = 4.3$ mkm, and after 60 days of moistening 6.87 mkm. During aging the number of defects increased 10%. With further humidification of the coating in the places of stress concentration (distortion of the pattern of interference fringes on the interferogram) cracks and

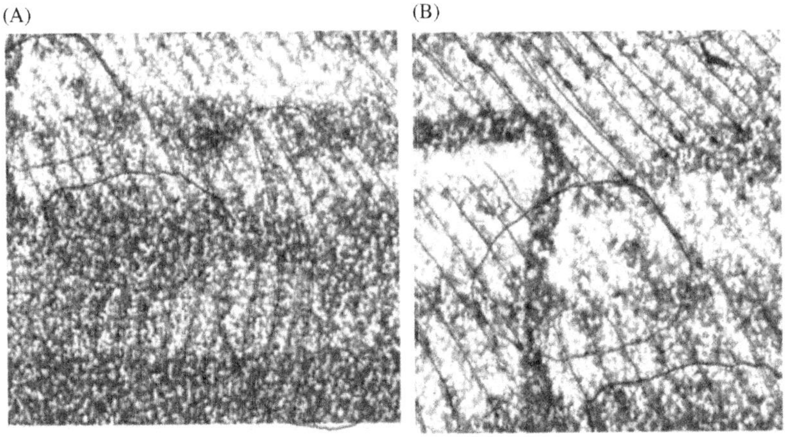

Figure 2.7 Areas of disturbance and distortion of the order of the interference. (A) D8-C8-C9-D9 and D9-C9-C10-D10 and (B) C8-B8-B9-C9 (paint surface MA-15, dynamic viscosity 26.0 Pa s and porosity of the substrate 32%).

Table 2.1 Change in the quality of the appearance of coatings in the process of freezing and thawing

Name of paint	Surface roughness, R_a (mkm)	Number of defects after curing	Number of defects/probability of fracture coatings (%)			
			Test cycles			
			5	10	13	15
Alkyd enamel PF-115	2.78	18	25/35.5	31/40.7	56/47.2	57/53.8
	3.7	29	38/31.0	56/100[a]	–	–
	5.1	35	43/32.4	77/100[a]	–	–
	5.6	54	67/33.4	80/100[a]	–	–
Oil paint MA-15	2.14	10	15/34.0	20/35.2	21/39.2	26/40.8
	2.18	20	23/35.1	26/38.1	30/40.0	33/41.3
	3.3	29	60/100[a]	–	–	–
	5.6	30	72/100[a]	–	–	–
	8.0	25.6	82/100[a]	–	–	–
Aqueous dispersion (facade)	2.4	40	44/39.6	47/45.8	56/46.8	64/48.2
	2.7	55	61/42.8	64/47.2	68/47.9	71/49.3
	3.4	56	67/43.2	70/47.6	74/48.0	77/49.8
	4.4	60	69/45.1	72/48.7	83/49.0	91/51.8

[a]There is peeling of coating.

Table 2.2 Change in the quality of appearance of coating in the process moistening and drying

Name of the ink composition	Surface roughness, R_a (mkm)	The number of defects after curing	Number of defects/probability of fracture coatings (%)				
			Test cycles				
			5	8	11	13	15
Alkyd enamel PF-115	3.2	31	36/48.2	42/57.6	50/58.3	59/100[a]	–
	4.0	32	37/48.8	47/100[a]	–	–	–
	5.8	39	48/49.7	59/100[a]	–	–	–
	6.2	44	59/50.2	60/100[a]	–	–	–
Oil paint MA-15	4.6	13	16/43.9	24/50.6	34/100[a]	–	–
	6.9	22	28/44.8	37/100[a]	–	–	–
	9.1	33	54/46.3	87/100[a]	–	–	–
Aqueous dispersion (facade)	2.55	50	60/34.3	64/36.8	67/39.8	70/44.0	72/46.8
	3.07	62	67/36.2	73/39.2	81/42.3	89/45.7	95/48.3
	3.29	68	73/36.8	81/39.6	86/42.6	92/46.3	98/48.7

[a]There is a peeling of coating.

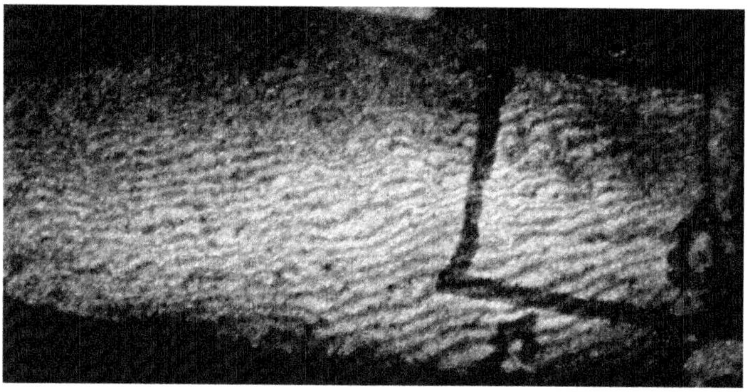

Figure 2.8 Interferogram coating based on paints MA-15 after 60 days of moistening (dynamic viscosity of 26.0 Pa s and porosity of substrate 28%).

peeling paint were seen. A fragment of the coating with a crack, which appeared after 70 days of moistening, is shown in Fig. 2.9.

The results indicate a relation between SSS and resistance of coatings with the quality of their appearance. This creates the possibility to control the lifetime of coatings by adjusting the rheological properties of paints, the degree of surface preparation to be coated, etc.

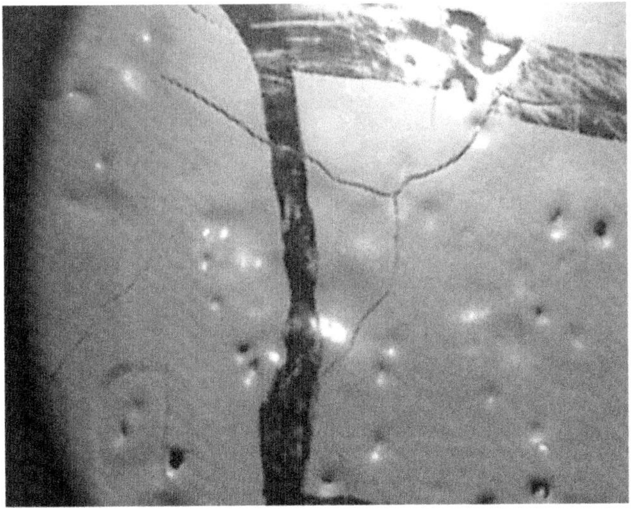

Figure 2.9 A crack in the coating based on paint MA-15 after 70 days of moistening (dynamic viscosity of 26.0 Pa s and porosity of substrate 28%).

2.3 Stress state of the coatings, depending on the operational factors

To assess solidity of the finishing layer it is necessary to study the stress-state coatings [5,6]. With respect to the contact area coating of the substrate, the internal stresses can act in different directions. Specialists distinguish the directed tangentially and the directed normal stresses.

At a rough substrate (the surface of the plaster, concrete, asbestos cement) the surface protrusion, to some extent, "absorbs" the internal stresses. By increasing the thickness of the coating, when the coating thickness exceeds the height of the projections, the projections themselves are not able to compensate for the effects of stress, which, in some cases, also lead to cracking of the coatings.

The relationship between the stress depending on the height of the projections of a rough surface β and the coating thickness h is given by [13,14]

$$\sigma^{uu} = \left[1 - \frac{k\beta}{h+\beta}\right]\sigma \tag{2.3}$$

where $\sigma$is the stress on the smooth surface.

Different relationships between the stress on a smooth and rough surfaces depend on the numerical values of the coefficient k and the thickness of the coating h. Taking into account the discrete nature of the contact, we need to consider the accounting pores, not filled with a paint.

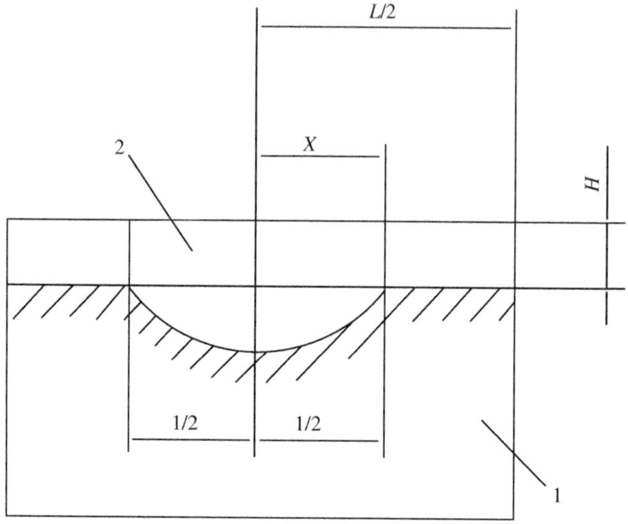

Figure 2.10 Calculated coating model.

The colorful compositions were applied in two layers with intermediate drying for 20 minutes.

In total, we made 50 measurements. The coating thickness and pores diameter of the substrate was measured with a micrometer. The distance between the cracks was measured with a ruler. The stress value was obtained by calculation. The distances between the cracks were measured after 90 days hardening of coatings.

The coating stress strain properties estimation was carried out with the help of a tensile machine IR5057-50 with the samples after 28 days of air and dry curing. This method is based on the sample stretching until it ruptures at a deformation speed of 1 mm/min. The $1 \times 1 \times 5$ cm samples were fixed in the clips of the tensile machine so that their longitudinal axis was in the direction of stretching, and the force was applied equally all over the sample section. The tests were carried out at the temperature of 20°C and relative air humidity of 60%. The ultimate tensile strength estimation was carried out for no less than four samples of each compound. The ultimate tensile strength R_{kog} for each sample was calculated by

$$R_{kog} = \frac{F_{Pi}}{S_{Oi}} \tag{2.4}$$

where F_{Pi} is the stretching loading at the time of a rupture, N, and S_{Oi} is the initial cross-sectional area of a sample, m^2.

Fig. 2.10 shows a calculation model of the coating (layer 2). For coating strip a single width, the snugly fit to the surfaces, the differential equation of equilibrium has the

$$\frac{d^2U}{dx^2} - n^2U = -n^2\varepsilon x \tag{2.5}$$

where $n = \sqrt{G/HE}$, $U(x)$ is the unknown displacement uniaxial strained coating, E is the modulus of elasticity of the coating and$G$ is the shear modulus of the coating.

For the coating over the pores (where there is no interaction with the surface) the differential equation equilibrium takes the form

$$U(x) = 0 \tag{2.6}$$

Thus, the problem reduces to the solution of two differential equations: Eq. (2.5) for the coating over pores ($0 < x < 1/2$), for the coating portion in contact with the surface ($1/2 < x < L/2$). The solution of these equations has the form

$$U_1(x) = C_1x + D_1 \tag{2.7}$$

1. The integration constants C_1, D_1, C_2, D_2 are determined from the boundary conditions and
2. the condition of symmetry with respect to movement of the center of the pores

$$\text{at } x = 0\ U_1 = 0 \tag{2.8}$$

3. The condition of continuity of movement and deformation of the pores on the border

$$\text{at } x = 1/2\ U_1 = U_2U'_1 = U'_2 \tag{2.9}$$

4. and conditions in the absence of stress in fracture

$$\text{at } x = L/2,\ U_2 = O \tag{2.10}$$

Substituting the values of the constants in Eq. (2.7) we can obtain a final solution of the original differential equations:

$$U_1(x) = \varepsilon x[1 - 1/ch(\beta - \alpha) + \alpha sh(\beta - \alpha)] \tag{2.11}$$

at $0 < x < l/2$

$$U_2(x) = \varepsilon[1 + [ch(nx - \alpha) + ash(nx - \alpha)]/[ch(\beta - \alpha) + ash(\beta - \alpha)] \tag{2.12}$$

at $l/2 < x < L/2$ where $\alpha = nl/2; \beta = nL/2;$

Over pores the relative deformations and corresponding stress $\sigma = U$ a finishing layer is constant:

$$U_1(x) = \varepsilon[1 + 1/ch(\beta - \alpha) + ash(\beta - \alpha)] \tag{2.13}$$

Outside of the pores relative deformations change by the law

$$U_2(x) = \varepsilon[1 + [ch(nx - \alpha) + ash(nx - \alpha)]/[ch(\beta - \alpha) + ash(\beta - \alpha)] \tag{2.14}$$

The cracks in the finishing layer will appear if the deformation layer exceeds the limit deformation $\varepsilon_{npe\partial}$:

$$\varepsilon[1 + 1/ch(\beta - \alpha) + ash(\beta - \alpha)] \geq \varepsilon_{npe\partial} \tag{2.15}$$

From this condition we can obtain the estimated distance between the shrinkage cracks in the finishing layer:

$$L = 1 + 2/n \ln\{\varepsilon/(\varepsilon - \varepsilon_{npe\partial})(1 + \alpha)x1 + \sqrt{1 - (1 - \alpha^2)}(\varepsilon - \varepsilon_{npe\partial})/\varepsilon^2 \tag{2.16}$$

In accordance with the mentioned above equations, the values of stress in the coatings depending on the pores size and coating thickness on polyvinyl acetate cement PVAC paint have been calculated. The results of calculations and studies are summarized in Table 2.3.

Analysis of the experimental data (Table 2.3) suggests that when coating thickness up to 0.09 cm, coating cracking is not observed. This is also confirmed by the calculated data according to Eq. (2.16). The stress in the coating at pores is

Table 2.3 Values of the normal stresses in polyvinyl acetate cement PVAC coating

Coating thickness, *H* (cm)	**Pores diameter (cm)**	**Value of the stress (MPa)**	
		Coating above pores	**Coating in contact with the substrate**
1. The appearance of cracks is not expected.			
0.09	0.3	0.585	0.519
0.06	0.2	0.585	0.519
	0.01	0.585	0.519
	0.05	0.585	0.519
	0.3	0.54	0.501
	0.2	0.54	0.501
	0.01	0.54	0.501
	0.05	0.54	0.501
2. The appearance of cracks is predicted.			
0.18	0.2	0.657	0.069/2.094
0.225	0.1	0.657	0.061/2.084
	0.05	0.657	0.059/2.08
	0.0	–	0.0582/2.079
	0.2	0.657	0.041/2.056
	0.1	0.657	0.033/2.046
	0.05	0.657	0.031/2.043
	0.0	–	0.039/2.042

Notes: The values of the stress are given for $x = 7$ cm at length of contact $l = 16$ cm. Above the line indicated values of the stress for $x = 1$ cm; below the line the distance between the shrinkage cracks, cm.

constant, at coating thickness $H = 0.06$ cm it is 0.54 MPa and at coating thickness $H = 0.09$ cm it is 0.585 MPa. The stress in the coating in contact with the substrate is 0.519 and 0.501 MPa at $x = 7$ cm at length of contact 16 cm.

In the absence of cracks in the coating the stress dependence of the pores diameter in the contact zone does not appear.

When the coating thickness is greater than 0.09 cm, relative deformations of the coating exceed the value of maximum elongation (in accordance with Eq. (2.16)), which is confirmed by the experimental data. In this case, the stresses in the coating over pores (no filled paint) are 0.657 MPa. In the coating area in contact with the substrate observed dependence of stress on the pores diameter (Table 2.3) is seen. Thus, when the coating thickness $H = 0.225$ cm, the magnitude of internal normal stresses at pores diameter $d = 0.2$ cm, at the contact boundary it is 0.041 MPa and when the diameter of the pores $d = 0{,}05$ cm, it is 0.0315 MPa. Dependence of the stress on the pores diameter is more pronounced when reducing the thickness of the coating. Thus, at coating thickness $H = 0.18$ cm, the values of stresses are 0.069 and 0.0591 MPa, respectively.

The presence of pores in the contact area coating with the cement substrate contributes to a more inhomogeneous SSS in comparison with the smooth, without pores substrate. In order to improve the crack resistance and, hence, service durability of coatings, we have to seek technological methods to create the cement substrates that are characterized by small pores uniformly distributed. The approach can be applied to the development of the coating quality control. The given technique can be used in the development of protective properties of paint and varnish coatings for various surfaces and operation conditions, and also for solving the problem of forecasting protective properties of coatings.

The temperature–time dependence of the strength of paint and varnish materials can be described by the Zhurkov equation [15–18]:

$$\iota = \iota_o \exp\,[U_o - \gamma\sigma)/RT \tag{2.17}$$

where γ is the structural-sensitive factor characterizing overstrain of bonds in the structure of the material, $U_o$is the activation energy of the process of destruction, $R$is the universal gas constant and$T$ is the absolute temperature.

The values of the activation energy of the fracture process and the structure-sensitive factor for polyvinyl acetate-cement PVAC and organosilicon KO-168 coatings were calculated.

Figs. 2.11 and 2.12 show the experimental dependence of the long-term cohesive strength of coatings from the value of stresses and temperatures for the coatings under study in semilogarithmic coordinates. The values of the structure-sensitive factor and the activation energy of the process of destruction of coatings are given in Table 2.4.

Analysis of the data in Table 2.4 shows that the activation energy of cohesive destruction of coatings decreases with increasing stresses acting on the coatings. A higher value of the activation energy and a lower value of the structurally

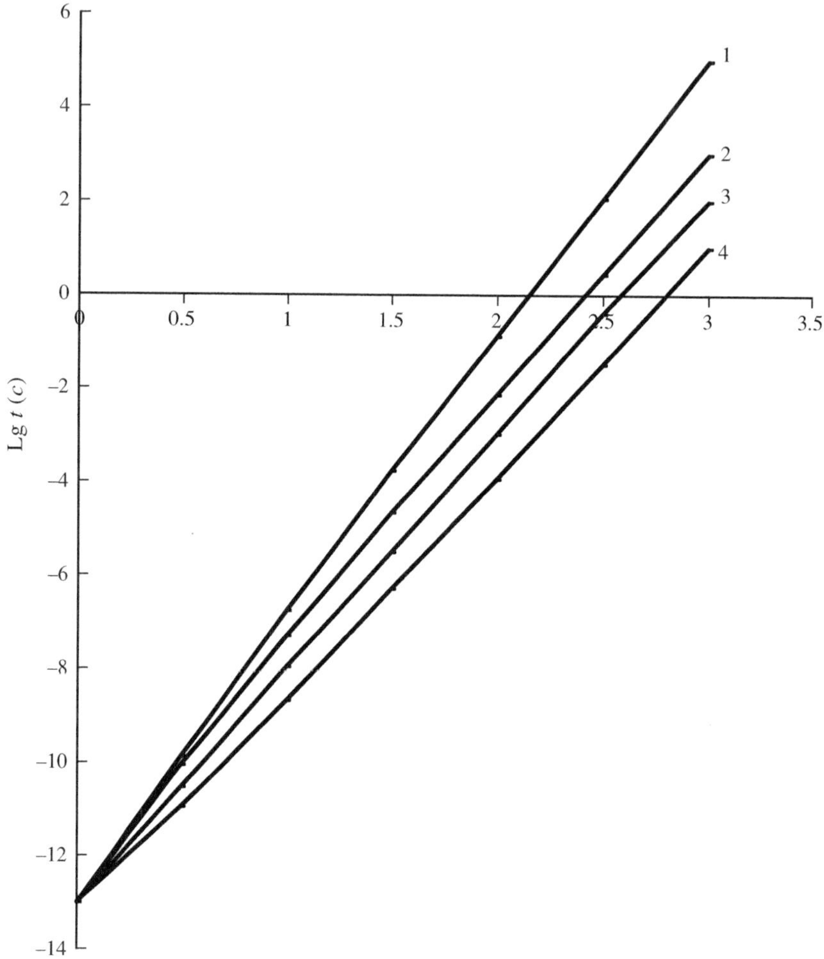

Figure 2.11 Temperature dependence of Lg τ -(σ)
1—coating of PVAC, $\sigma = 0.166$ MPa
2—coating of PVAC, $\sigma = 0.124$ MPa
3—coating of KO-168, $\sigma = 0.144$ MPa
4—coating of KO-168, $\sigma = 0.17$ MPa.

sensitive factor indicate high strength of polyvinyl acetate cements compared with organosilicon coatings.

The duration of preservation of cohesive strength of coatings during operation is also determined by the resistance to periodic effects of environmental factors: wetting-drying, freezing-thawing, etc. In this connection, the influence of humidification on the change in the duration of cohesive strength was assessed. To this end, stretched coating samples were subjected to sprinkling. The results of the tests are shown in Fig. 2.13.

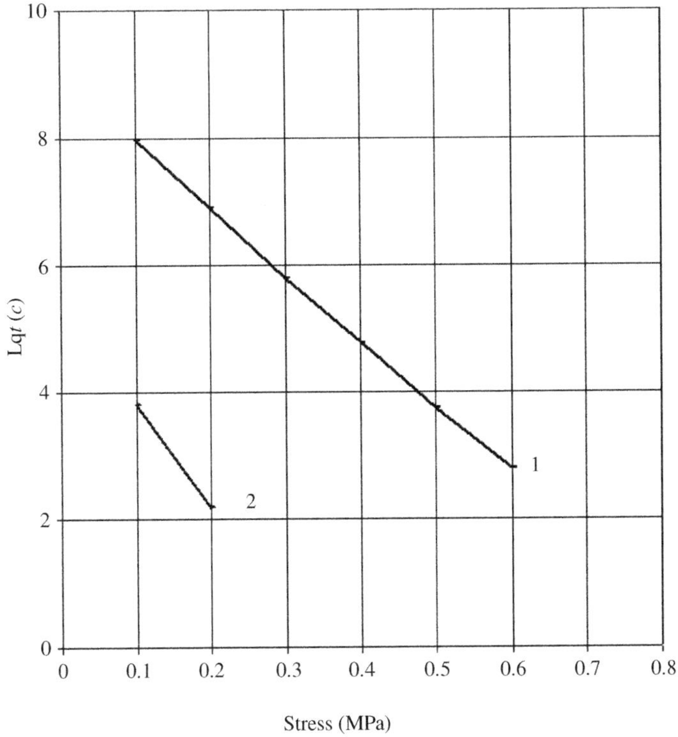

Figure 2.12 Dependence of the long-term strength of coatings from stresses
1—coating of PVAC
2—coating of KO-168.

Table 2.4 Values of U and γ of PVAC and organosilicon KO-168 coatings

Type of coating	Activation energy (kJ/mol)	Stresses (MPa)	U_o	γ
Polyvinyl acetate cement	116.09 114.42 111.66	0.24 0.3 0.4	122.81	27.98
Silicone KO-168	97.47 95.24	0.14 0.17	102.53	42.86

Note: For coatings of PVAC and KO-168 $\tau_o = 10^{-13}$ s.

Consider the condition of brittle cracking of polymer coatings under investigation under the action of internal stresses [15,16]. In the case of brittle failure, the cracking condition has the form

$$\sigma \geq 0.5R \tag{2.18}$$

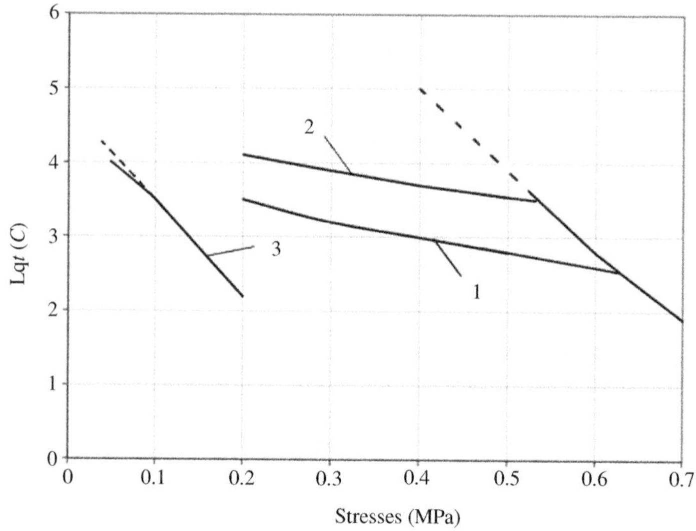

Figure 2.13 Dependence of the long-term strength of coatings in the process of humidification
1—coating of PVAC
2—coating of PVAC with surface hydrophobization
3—coating of KO-168.

Table 2.5 Physicomechanical properties of coatings after moistening

Name of coating	Internal stresses (MPa)	Cohesive at strength of coatings (MPa)	σ/R	Time of occurrence cracks (day)
PVAC	0.08	0.38	0.21	Not observed
Polymer-lime	0.062	0.13	0.47	27
Lime	0.075	0.09	0.9	30

Consider the internal stresses and the strength of coatings in the aging process in the example of humidification. As can be seen, at the first instant humidification is observed a sharp increase internal stresses, and then their slow decrease. For example, for PVAC coatings after 20 days of moistening stabilization of internal stresses at level $\sigma = 0.08$ MPa was observed, and stabilization of internal stresses in polymer lime coatings was observed after 18 days of humidification at level $\sigma = 0.062$ MPa. The values of short-term strength after 60 days of moistening are given in Table 2.5.

The polyvinyl acetate cement coating after 2 months of moistening did not have cracks, and the ratio $\sigma/R$for such coatings did not exceed 0.21.

References

[1] I.V. Volkov, Construction Surface Deformation Determination, Invention Certificate No 1269635 (July 8, 1986).

[2] Holographic Non-Destructive Research (Translated from English under editorship of R. K. Erf, V.A. Karasyov). Mashinostroyeniye, Moscow, 1979, 448 p.

[3] N. Abramson, Praktical interpretation of holographik interferograms, Optik 37 (№3) (1972) 337–346.

[4] Holographic Non-Destructive Testing (Translation from English edited By R.K. ERF, V.A. Karaseva). Machine Building, Moscow, 1979, 448 c.

[5] G.S. Landsberg, Optics Text, Science, Moscow, 1976, p. 207.

[6] Yu. I. Ostrovsky, V.P. Sabatino, V.V. Yakovlev, Holographic interference methods of deformation measurement, Moscow, Science. GL. ed. Fiz.-Mat. lit. (1988). 248 p.

[7] Butters, J. Holography and Its Application Text, John. Butters. M.: Science, 1973-720C.

[8] A.G. Kozachok, Holographic Research Methods in Experimental Mechanics, Mechanical Engineering, Moscow, 1984. 176 p.

[9] V.I. Loganina, Y.P. Skachkov, The application of the holographic method for evaluation of a stress deformation state of cement paint coatings, Int. J. Appl. Eng. Res. 11 (14) (2016) 8377–8378.

[10] L.P. Orentlicher, V.I. Loganina, Protective Decorative Coatings for Concrete and Stone Buildings. Study Guide, Stroyizdat, Moscow, 1992.

[11] V.I. Loganina, The influence of surface quality of coatings on their deformation properties, Contemp. Eng. Sci. 7 (36) (2014) 1935–1941. Available from: https://doi.org/10.12988/ces.2014.411241. HIKARI Ltd, www.m-hikari.com.

[12] V.I. Loganina, Maintenance of quality of paint and varnish coverings of building products and designs, Contemp. Eng. Sci. 7 (36) (2014) 1943–1947. Available from: https://doi.org/10.12988/ces.2014.411243. HIKARI Ltd, www.m-hikari.com.

[13] A.D. Zimon, Adhesion of Films and Coatings, Chemistry, Moscow, 1977. 351 p.

[14] G.I. Gorchakov, L.P. Orentlicher, V.I. Savin, et al., Composition, Structure and Properties of Cement Concretes, Stroyizdat, Moscow, 1976. 144 p.

[15] S.N. Zhurkov, Kinetic concept of the strength of solids, Zhurkov-Vestnik//USSR Acad. Sci. (3) (1968) 46–52.

[16] V.R. Regel, A.I. Slutsker, E.G. Tomashevsky, Kinetic Nature of Strength of Solids, The Science, Moscow, 1974. 560 p.

[17] Yakov Il'ich Frenkel, Kinetic Theory of Liquids, Science, Frenkel-Leningrad, 1979. 592 p.

[18] N.A. Shtyrev, The Equation of State and the Structural-Energy Kinetic Law Deformed Solid, The Energy of Durability, 2013, p. 4, http: //energydurability.com.5.

Regularities of cracking protective-decorative coatings

3

3.1 Method for assessing the crack resistance of paint and varnish coatings

In this chapter we assess the cracking of polymer coatings using a technique based on the ratio between the crack length, the Vickers indenter print and the fracture toughness. This technique has been successfully used to assess the fracture toughness (K_{1c}) of ceramics [1].

In this method, the value of the stress intensity factor K_{1c} is determined from the length of the radial cracks formed in brittle materials from the corners of the Vickers indentation. To obtain a semiempirical dependence of the length of radial cracks on crack resistance and hardness, the approach proposed by A. Evans and E. Charles is used.

Dependence "normalized crack resistance $(K_{1c}/H\alpha^{1/2})\cdot(H/E)^{1/2}$-normalized length of radial cracks C/α" is described by the following equation:

$$(K_{1c}/H\alpha^{1/2})\times(H/E)^{1/2}=A\times(C/\alpha)^{-B} \tag{3.1}$$

where H is Vickers hardness and A and B are empirical coefficients. From the expression (3.1) we obtain:

$$K_{1c}=AH\alpha^{1/2}(E/H)^{1/2}(C/\alpha)^{-B} \tag{3.2}$$

If the value of hardness does not depend on the load on the indenter, then Eq. (3.2) can be written in the form:

$$K_{1c}=\text{const}\left(E/H\right)^{1/2}P/C^{B} \tag{3.3}$$

where P is the load on the Vickers indenter.

The most common semi-empirical equation of this type is

$$K_{1c}=0.028H\alpha^{1/2}\left(E/H\right)^{0.5}(C/\alpha)^{-1.5} \tag{3.4}$$

where $H=$ Vickers hardness

$E=$ The modulus of elasticity
$C=$ Half-length of radial cracks
$a=$ Half-length of the diagonal of the print

Increasing the Durability of Paint and Varnish Coatings in Building Products and Construction.
DOI: https://doi.org/10.1016/B978-0-12-817046-5.00003-6

The critical coefficient of intensity of tensions was determined by a formula:

$$K_{1c} = 0.028\, Ha^{0.5}(E/H)^{0.5}(C/a)^{-1.5} \tag{3.5}$$

where $H =$ hardness by Vickers

$P =$ loading on indents
$C =$ semilength of radial cracks
$a =$ semilength of print diagonal

Since colorful structures in work polyvinyl acetate cement PVAC and polymer-calcareous paints were applied, and as a substrate-cement and sand solution. After curing the painted solution exemplars were subjected to alternate freezing and thawing, and also to humidification and thermal aging. During tests with the help of the Vickers indenter we measured the print diameter and the length of the radial cracks formed on both sides of the print. Hardness by Vickers was calculated using the formula:

$$\mathbf{H} = \frac{2P}{d^2} \sin \frac{\alpha}{2} \tag{3.6}$$

where d is the diameter of a print and is the angle at the indenter top.

α- Angle at indenter top.

The dependence of the size of the semidiagonal of the imprint on the load on the indenter is described by the equation:

$$a = A \times P^{0.5} \tag{3.7}$$

The exponent is 0.5, as it should be when the hardness of the material does not depend on the magnitude of the load.

The dependence of the length of radial cracks on the load is fairly accurately approximated by an equation of the form:

$$C = B \times P^{0,67} \tag{3.8}$$

To ensure that the critical value of the intensity factor K_{1c}, measured by the method of indentation, does not depend on the load, the exponent must be equal to 2/3 ($\approx$ 0.67). The results of previous research work show that most protective and decorative coatings have a fragile character of destruction, which gives grounds to apply this technique to assess the crack resistance of paint and varnish coatings. From the experimental data obtained, a correlation was established between the semidiagonal of the imprint and the load on the indenter P, which is described by expression (3.9).

For the PVAC coatings tested after curing obtained dependence

$$a = 0.087P^{0.5} \tag{3.9}$$

For polymer-calcareous coatings

$$a = 0.124P^{0.5} \tag{3.10}$$

For coatings based on oil paint

$$a = 0.097P^{0.5} \tag{3.11}$$

For coatings KO-168

$$a = 0.141P^{0.5} \tag{3.12}$$

Eqs. (3.9)–(3.12) indicate that the exponent $a = 0.5$, as should be the case if the hardness is independent of the load in accordance with Eq. (3.7). After a certain duration of thermo stating of the coatings, a decrease in the coefficient A in Eq. (3.7) is seen, which indicates an increase in their hardness. Thus, for example, after 100 hours of thermo stating the following dependences were obtained:

For PVAC coatings

$$a = 0.065P^{0.5} \tag{3.13}$$

For polymer-calcareous coatings

$$a = 0.085P^{0.5} \tag{3.14}$$

For coatings based on oil paint

$$a = 0.073P^{0.5} \tag{3.15}$$

Analysis of the obtained results shows that the dependence of the half-length of radial cracks C on the load P in the coatings is approximated by the equations:

For polymer-calcareous coatings

$$C = 0.0327P^{0.67} \tag{3.16}$$

For PVAC coatings

$$C = 0.042P^{0.67} \tag{3.17}$$

For coatings KO-168

$$C = 0.252P^{0.67} \tag{3.18}$$

The results of experimental studies indicate that in the process of aging protective decorative coatings of external walls of buildings there is a change in the mechanism of their destruction from elastic-plastic to brittle, that is, "embrittlement" of coatings is observed. According to linear mechanics, fracture cracking of coatings occurs if

$$K_1 \geq K_{1c} \tag{3.19}$$

$K_1 =$ stress intensity factor
$K_{1c} =$ the critical value of the stress intensity factor
Test results are provided in Table 3.1.

It should be noted that the properties of protective-decorative coatings are not uniform along the stretch (length). In connection with this, the question of place of identification intensity coefficient of the coatings is important. When choosing the location of the stress intensity factor for the coatings, G.I. Gorchakov stress calculations for layered systems were used [2–4]. This design model represents a thin finishing layer (up to 1–2 cm thick) on a thick base in the form of a panel or block. In this case, the finishing layer does not have a significant mechanical effect on the substrate, and the monolithic finish depends on the difference in shrinkage (or temperature) deformation of the substrate (ε_2) and the finishing layer (ε_1):

$$\varepsilon = \varepsilon_2 - \varepsilon_1 \tag{3.20}$$

The distribution of stresses in the process of thermal aging of coatings of cement concretes was investigated. As substrates we used materials characterized by various values of coefficient of linear thermal expansion (CLTE) heavy concrete, haydite concrete. Values of KTLD for the considered materials are presented in Table 3.1. The sizes of the exemplars of a substrate are $4 \times 4 \times 16$ cm^3. The results of the calculations are presented in Figs. 3.1 and 3.2.

Table 3.1 Values of coefficients of temperature linear dilatation

№	Name of material	Values CLTE $\times 10^{-6}$, 1/degree
1	PVAC coating	8.43
2	Polymer-calcareous coatings	3.47
3	Haydite concrete of structure (on volume) 1:1.5:1.5	6.6
6	Heavy concrete	10

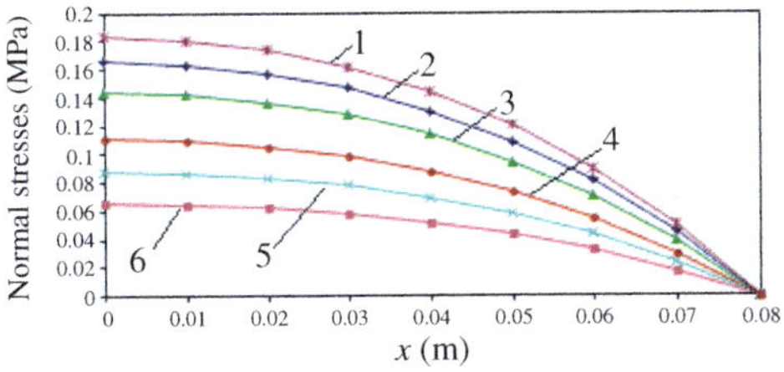

1—PVAC thermal aging on the heavy concrete
2—PVAC after curing
3—PVAC thermal aging on the lightweight concrete
4—polymer-calcareous on the heavy concrete
5—polymer-calcareous on the lightweight concrete
6—polymer-calcareous coatings after curing

Figure 3.1 Distribution of normal tensions on contact extent.

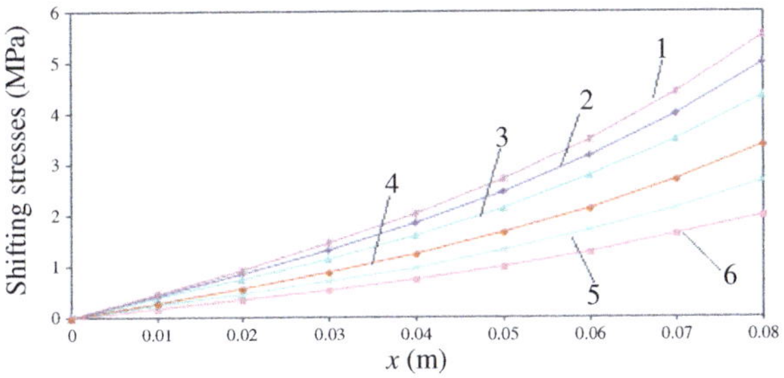

1—PVAC thermal aging on the heavy concrete
2—PVAC after curing
3—PVAC thermal aging on the lightweight concrete
4—polymer-calcareous on the heavy concrete
5—polymer-calcareous on the lightweight concrete
6—polymer-calcareous coatings after curing

Figure 3.2 Distribution of the shifting tension on contact extent.

In Fig. 3.1 distribution of normal tensions depending on the extent of contact of coating with a substrate is presented. The results show that the greatest tension after curing for PVAC coatings is $\sigma = 0.164$ MPa. For the polymer-calcareous coatings this value is $\sigma = 0.066$ MPa.

After curing a temperature increase to 60°C leads to increase in size of normal tensions in coatings. Thus for PVAC coating on a substrate from heavy concrete the size of normal tensions is $\sigma = 0.183$ MPa, and for a polymer-limy coating respectively it is $\sigma = 0.111$ MPa.

In Fig. 3.2 distribution of the shifting tensions depending on the extent of contact of a coating with a substrate is presented. The results show that the greatest tension is observed in coatings on a substrate from heavy concrete. After curing of coatings the shifting tension for PVAC coatings is $\tau = 5 \times 10^{-3}$ MPa. For the polymer-limy coatings this value is $\tau = 2 \times 10^{-3}$ MPa. After thermo aging there is an increase in the size of the shifting tensions. Thus for PVAC coatings on a substrate from heavy concrete this value is $\tau = 5.53 \times 10^{-3}$ MPa, and for polymer-calcareous coatings it is $\tau = 3.4 \times 10^{-3}$ MPa.

From the above, it is possible to conclude that the assessment of crack resistance of a protective decorative coating by Vikkers's method involves two stages:

1. assessment of crack resistance on an average section characterized by the greatest pulling tension;
2. assessment of crack resistance on an extreme section as the shifting tensions increase with increased extent of finishing layer.

This approach will help choose materials with crack resistance of coatings more reasonably.

These results justify the need for developing of compounding of colorful structures and carrying out research using technique of an assessment of crack resistance of coatings according to the offered scheme. It will allow to predict more reasonably firmness of coatings, and also to optimize finishing structures for the purpose of receiving coatings with a complex of the given properties.

3.2 Regularities of cracking protective and decorative coatings in the aging process

The results of experimental studies indicate that, in the process of aging protective decorative coatings of external walls of buildings there is a change in the mechanism of their destruction from elastic-plastic to brittle, i.e."embrittlement" of coatings is observed (Table 3.2).

It was established that in PVAC and the polymer-calcareous coatings on a solution substrate the "embrittlement" occurs after a particular duration of impact of alternate freezing and thawing. Cracks in coatings at cave-in of an indenter Vickers appear only after 15–20 testing cycles. Value of critical coefficient of intensity of tensions of PVAC coating is equal $K_{1c} = 0.088$ MH/m$^{3/2}$, and for polymer calcareous coating $K_{1c} = 0.069$ MH/m$^{3/2}$.

Introduction into a compounding of PVAC paint the fibrous micro excipient (asbestos) increases crack resistance of coatings. Thus even after 20 cycles of alternate freezing thawing the "embrittlement" of coating is not observed. The comparative analysis of data shows that at the same intensity of influences of the environment coatings with a fibrous micro excipient possess with smaller value of coefficient of intensity of tensions, after 8 cycles of alternate freezing thawing K_{1c}(PVAC) = 0.078 MH/m$^{3/2}$, and K_{1c}(PVAC with 1% of asbestos) = 0.073 MH/m$^{3/2}$.

Table 3.2 Parameters of a crack for the formation of a protective decorative coating

Type of coatings	Type of influence	Loading P, H	Hardness of coating, H, H/mm^2	The relation of crack semi-length C to the size of semi-diagonal of print	Coefficient of intensity of tension, K_{1c}, MH/m$^{3/2}$
PVAC	After curing	47.39	61	1	0.06
	3 cycles of freezing- thawing		137	1	0.075
	8 cycles of freezing-thawing		164	1	0.078
	15 cycles of freezing-thawing		179	1.2	0.088[a]
	Humidification 15 days		49	1	0.058
	Humidification 30 days		37	1	0.054
	Thermoaging 100 h		85	1	0.065
	Thermoaging 200 h		104	1	0.068
Polymer-calcareous	Aftercuring	47.39	27	1	0.044
	3 cycles of freezing-thawing		45	1	0.05
	20 cycles of freezing-thawing		70	1.54	0.069[a]
	Humidification 15 days		23	1	0.055
	Humidification 30 days		16	1	0.040
	Thermoaging 100 h		55	1	0.053
	Thermoaging 200 h		62	1	0.0546
PVAC(1% asbestosa by weight of cement)	After curing	47.39	44	1	0.055
	3 cycles of freezing-thawing		130	1	0.073
	8 cycles of freezing-thawing		130	1	0.073
	20 cycles of freezing-thawing		133	1	0.074

[a]Critical coefficient of intensity of tensions.

Humidification of coatings leads to decrease of an elastic modulus and hardness of coatings that reduces a danger of crack formation at deformation of a wall construction. Humidification of coatings during 30 days does not cause crack fissuring of coatings. The coefficient of intensity of tensions of PVAC coating after curing is equal $K_{1c} = 0.06$ MH/m$^{3/2}$, and after humidification $K_{1c} = 0.054$ MH/m$^{3/2}$. Similar data are obtained and for the polymer-calcareous coatings.

At studying a thermo aging it was recorded that the increase of time of thermo aging leads to natural increase of value of coefficient of intensity of tensions. For example, after a thermo aging of polymer- calcareous coatings during 100 h increase of value of coefficient of intensity of tensions is observed from $K_{1c} = 0.044$ MH/m$^{3/2}$ (after curing) to $K_{1c} = 0.053$ MH/m$^{3/2}$, and after 200 hours value makes $K_{1c} = 0{,}0546$ MH/m$^{3/2}$.

Considering that properties of a protective decorative coatings are defined among other factors by properties of the painted construction and are heterogeneous on an extension, we follow-up carried out calculation of tensions arising in coatings as a result of influence of various factors according to a technique [5–8].

It will allow approaching to choice of materials research factors of increase of crack resistance of coatings more reasonably.

The conducted researches are justification for recommendations at developing of compounding of colorful structures, at carrying out of research works with use of technique of an assessment of crack resistance of coatings according to the offered scheme. It will allow to predict more reasonably firmness of coatings, and also to optimize finishing structures for the purpose of receiving coatings with a complex of the given properties.

To establish the connection between the fracture toughness characteristics of the coatings and the quality of their appearance in the process of the corrosive effect of the external environment, we carried out the following experiment. Colorful compositions were applied by brush on the cement substrates in two layers with intermediate drying for 24 hours. The following paints were used: alkyd enamel grade PF-115, oil paint MA-15, nitrocellulose enamel НЦ-123, acrylate paint class "universal" and acrylic water-dispersion (facade) paint. Different quality of the appearance of the coatings was created by changing the porosity of the substrate and the rheological properties of the paint During the tests, the samples were subjected to various types of corrosion attack, namely, alternating freezing-thawing according to the regime: 4 hours freezing at a temperature of −18°C, 20 hours of thawing and wetting-drying by the regime: 20 hours of moistening at room temperature and 4 hours of drying at a temperature of 60°C. During the experiment, the concentration of defects on a surface of the coating was also determined. The number of defects was determined on the surface area of 64 cm^2. The results of the studies are given in Tables 3.3 and 3.4.

It was found that during the test the cracks appear locally and are formed near the defects on the surface of the coating. In particular, surface cracks (visible to the naked eye) are appeared alter five test cycles on coatings based on MA-15 paint,

Table 3.3 Crack resistance of coatings depending on the quality of the ir appearance in the process of freezing-thawing

Name of paint	Type of corrosion attack	Number of defects	Surface roughness, R_a	Coefficient of intensity of stresses, K_I, MH/m$^{3/2}$
Alkyd enamel PF-115	After hardening	36	0.12	0.01561
		30	0.10	0.01035
		18	0.08	0.01002
	5 freezing-thawing cycles	39	0.47	0.01708
		32	0.36	0.01677
		25	0.23	0.01076
	10freeze-thawing cycles	There is a peeling of the coating		
		36	2.58	0.01913
		31	2.21	0.01846
	13 freeze-thawing cycles	There is a peeling of the coating		
		56	3.10	0.02082
	15 freeze-thawing cycles	57	3.26	0.02123
Oil paint MA-15	After hardening	29	0.23	0.01855
		20	0.18	0.01177
		10	0.14	0.01170
	5 freeze-thawing cycles	There is a peeling of the coating		
		23	0.59	0.02056
		15	0.40	0.01864
	10 freeze-thawing cycles	26	1.69	0.02653
		20	1.46	0.02014
	13 freeze-thawing cycles	30	1.83	0.02903
		21	1.63	0.02134
	15 freeze-thawing cycles	33	2.1	0.02985
		26	1.95	0.02461

(Continued)

Table 3.3 (Continued)

Name of paint	Type of corrosion attack	Number of defects	Surface roughness, R_a	Coefficient of intensity of stresses, K_I, MH/m$^{3/2}$
NitrocelluloseNC-123	After hardening	27	0.19	0.00986
		19	0.17	0.00984
		8	0.14	0.00824
	5 freeze-thawing cycles	30	0.52	0.01377
		21	0.48	0.01306
		12	0.16	0.01061
	10 freeze-thawing cycles	There is peeling of the coating		
		28	2.78	0.01672
		18	2.32	0.01543
	13 freeze-thawing cycles	31	2.90	0.01802
		24	2.51	0.01701
	15 freeze-thawing cycles	28	2.65	0.01925
Waterdispersive (facade)	After hardening	192	0.44	0.01688
		113	0.34	0.01481
		80	0.24	0.01308
	5 freeze-thawing cycles	229	0.89	0.02103
		135	0.76	0.01799
		94	0.70	0.01409
	10freeze-thawing cycles	232	3.01	0.02305
		141	2.55	0.01967
		97	1.85	0.01503
	13 freeze-thawing cycles	233	3.65	0.02636
		148	3.10	0.02013
		106	1.98	0.01723
	15freeze-thawing cycles	247	3.80	0.02752
		153	3.40	0.02432
		114	2.80	0.02056

Acrylate class “Universal”	After hardening	160	0.24	0.01283
		120	0.22	0.01193
		86	0.20	0.01137
	5 freeze-thawing cycles	210	0.74	0.01716
		134	0.60	0.01513
		108	0.55	0.01391
	10freeze-thaw cycles	215	2.40	0.02146
		150	1.77	0.01692
		117	1.44	0.01403
	13 freeze-thawing cycles	223	2.62	0.02342
		161	1.83	0.01863
		123	1.63	0.01608
	15 freeze-thawing cycles	228	3.30	0.25631
		176	2.31	0.02018
		129	1.96	0.01801

Table 3.4 Crack resistance of coatings depending on the quality of their appearancein the process of moistening-drying

Name of the paint composition	The change in coefficient of intensity of stresses K_I, MH/m$^{3/2}$after the test cycles					
	Alter hardening	**5 cycles**	**8 cycles**	**11 cycles**	**13 cycles**	**15 cycles**
Alkyd enamel PF-115	$\frac{0.01561}{0.58}$	$\frac{0.01804}{0.83}$	There is a peeling of the coating			
	$\frac{0.01035}{0.4}$	$\frac{0.01581}{0.58}$	There is a peeling of the coating			
	$\frac{0.01002}{0.21}$	$\frac{0.01264}{0.32}$	There is a peeling of the coating			
Oil paint MA-15	$\frac{0.01855}{0.8}$	$\frac{0.02004}{1.43}$	There is a peeling of the coating			
	$\frac{0.01177}{0.69}$	$\frac{0.01658}{0.94}$	There is a peeling of the coating			
	$\frac{0.01170}{0.46}$	$\frac{0.01342}{0.65}$	$\frac{0.01532}{0.93}$	There is a peeling of the coating		

Nitrocellulose NC-123	$\frac{0.00986}{0.78}$	$\frac{0.01564}{1.02}$	There is a cracking of the coating			
	$\frac{0.00934}{0.6}$	$\frac{0.01268}{0.79}$	$\frac{0.01586}{1.12}$	There is a cracking of the coating		
	$\frac{0.00824}{0.32}$	$\frac{0.01194}{0.44}$	$\frac{0.01302}{0.68}$	$\frac{0.01683}{1.18}$	$\frac{0.01968}{1.32}$	There is a cracking of the coating
Water-dispersive (facade)	$\frac{0.01488}{3.01}$	$\frac{0.01568}{3.42}$	$\frac{0.01932}{3.58}$	$\frac{0.02236}{3.72}$	$\frac{0.02393}{3.94}$	$\frac{0.02538}{4.16}$
	$\frac{0.01471}{2.55}$	$\frac{0.01496}{2.76}$	$\frac{0.01906}{2.96}$	$\frac{0.02198}{3.16}$	$\frac{0.02363}{3.28}$	$\frac{0.02506}{3.34}$
	$\frac{0.01308}{1.85}$	$\frac{0.01386}{1.94}$	$\frac{0.01896}{2.08}$	$\frac{0.02106}{2.34}$	$\frac{0.02186}{2.46}$	$\frac{0.02483}{2.61}$
Acrylate class “universal”	$\frac{0.01282}{2.4}$	$\frac{0.01462}{2.62}$	$\frac{0.01768}{3.12}$	$\frac{0.02068}{3.28}$	$\frac{0.02238}{3.38}$	$\frac{0.02368}{3.52}$
	$\frac{0.01193}{1.77}$	$\frac{0.01266}{1.94}$	$\frac{0.01701}{2.26}$	$\frac{0.02032}{2.43}$	$\frac{0.02113}{2.64}$	$\frac{0.02309}{2.83}$
	$\frac{0.01137}{1.51}$	$\frac{0.01204}{1.63}$	$\frac{0.01632}{1.84}$	$\frac{0.01958}{1.96}$	$\frac{0.02073}{2.18}$	$\frac{0.02298}{2.31}$

characterized by a roughness index $R_a = 0.23$ μm, and on the coating with a roughness index $R_a = 0.14$ μm after 15 test cycles . Similar regularities are also characteristics for other coatings (Table 3.3).

It is established that with an increase in the roughness of the coating surface, the value of the stress intensity coefficient is increased. So, for example, the surface roughness of the PF-115 paint coating is $R_a = 0.12$ μm, and coefficient of intensity of stresses $K_1 = 0.01561$ MH/m$^{3/2}$, at the roughness of the coating surface $R_a = 0.08$ μm and $K_1 = 0.01002$ MH/m$^{3/2}$. Similar patterns are also characteristic for other types of coatings.

The results (Table 3.4) show that the nature of the destruction of the coatings during the corrosive action of the medium is not the same. So, coatings based on oil and alkyd paint are characterized by peeling, and coatings based on acrylate class universal, nitrocellulose and water dispersion paint are characterized by cracking. Regardless of the type of coating and the corrosive effect of the medium, there is an increase in the roughness of the coating surface and coefficient of intensity of stresses.

3.3 Effect of the porosity of the cement substrate on the crack resistance of protective and decorative coatings

The operational stability of protective-decorative coatings of the outer walls of buildings is significantly influenced by processes occurring both in the coating itself and at the interface of the "substrate-coating" contact [9,10]. The strength of the adhesion of protective-decorative coatings to the concrete substrate depends significantly on the quality of the substrate. The quality of the substrate, first of all, is understood as its macro- and microstructure, the degree of its homogeneity, providing the desired solidity contact layer, its density and porosity. Features of the porous substrate, such as cement concrete, mortar, etc., have a significant effect on the formation of the structure and properties of the coatings applied. The analysis of the results (Table 3.5) indicates, that the porosity of the substrate has a great influence on the nature of the destruction of the protective and decorative coating [11]. At the same time, it should be noted that the nature of the destruction, for example, of PVAC coatings, has a significant difference from polymer-based coatings. Thus, for example, an increase in the porosity of the substrate from 20% to 28% leads to a reduction in the crack resistance of polymer-calcareous coatings. The appearance of cracks in indentation of the Vickers indenter in a polymer-calcareous coating on substrates with a porosity of $P = 20\%$ is observed after 20 cycles of freezing-thawing. The critical coefficient of intensity of stresses is $K_{1c} = 0.06$ MH/m$^{3/2}$. At porosity of the substrate, $P = 28\%$, the appearance of cracks in indentation of the Vickers indenter occurs in the coatings after 14 cycles of freezing-thawing. This is explained by the appearance of a more inhomogeneous stress state in the coating. For PVAC

Table 3.5 Parameters crack education of protective-decorative coatings depending on the porosity of the substrate

Name of coating	Impact type	Hardness H, N/mm^2	The ratio of the half-length of the crack C to the size of the semi-diagonal of the imprint a	Coefficient of intensity of stresses, K_I, MH/m$^{3/2}$
1	2	3	4	5
PVAC on glass	After hardening	37	1	0.051
	5 freeze-thaw cycles	76	1	0.072
	8 freeze-thawing cycles	85	1.27	0.083[a]
($P = 20\%$) PVAC	After hardening	85	1	0.065
	5 freeze-thawing cycles	125	1	0.08
	10 freeze-thawing cycles	140	1	0.084
	11 freeze-thawing cycles	180	1.4	0.088[a]
($P = 28\%$) PVAC	After hardening	75	1	0.065
	5 freeze-thaw cycles	80	1	0.066
	10 freeze-thawing cycles	102	1	0.08
	15 freeze-thawing cycles	174	1.23	0.083[a]
PVAC on a brick substrate ($P = 40\%$)	After hardening	73	1	0.077
	5 freeze-thawing cycles	176	1.3	0.085[a]
($P = 20\%$) polymer-calcareous	hardening	21.7	1	0.024
	5 freeze-thawing cycles	47.6	1	0.05
	10 freeze-thaw cycles	52.2	1	0.052
	20 freeze-thaw cycles	60.8	1.54	0.06[a]
($P = 28\%$) polymer-calcareous	After hardening	27	1	0.029
	5 freeze-thawing cycles	38	1	0.04
	14 freeze-thawing cycles	57	1.2	0.056[a]
($P = 20\%$) Acrylate class "universal"	After hardening	61	1	0.027
	curing on preliminary putty surfaces	22	1	0.004

[a]Critical coefficient of intensity of stresses.

coatings it is characteristic that the fracture toughness increases with increasing porosity of the substrate from 20% to 28%. The critical value of coefficient of intensity of stresses for PVAC coatings in substrate porosity of 20% is $K_{1c} = 0.088$ MH/m$^{3/2}$. In this case, the appearance of cracks in the coating during the introduction of the Vickers indenter was recorded after 11 cycles freezing-thawing. With a substrate porosity of 28%, the appearance of cracks is observed only after 15 cycles of freezing-thawing. However, the change in coefficient of intensity of stresses for PVAC coatings is extreme. So, for example, when using as a substrate of brick samples with a porosity of $P = 40\%$, the fracture toughness of PVAC coatings is significantly reduced. Thus, the presence of cracks at indentation of Vickers indenter is observed after 5 cycles of freezing thawing.

Preliminary preparation of the substrate surface has a significant effect on the crack resistance of protective and decorative coatings. So, for example, priming the surface of the substrate leads to an increase in the crack resistance of the coatings. At the same time, the appearance of cracks with indentation of the Vickers indenter in the PVAC coating on a substrate with a porosity of $P = 20\%$ was observed after 15 cycles of alternating freezing-thawing. For some types of coatings PF-115 and water dispersive acrylate paint of the "universal" class, it is very important in advance application of putty to the surface before staining to reduce the porosity of the substrate. This leads to an increase in the crack resistance of these coatings. Thus, coefficient of intensity of stresses of the acrylate coating applied to the surface of the substrate after its preliminary preparation was = 0.004 MH/m$^{3/2}$, while without preparation = 0.027 MH/m$^{3/2}$.

It was found, that in coatings "embrittlement" occurs after a certain time of moistening. Thus, cracks in the coatings with indentation of the Vickers indenter appear on the coating of PF-266 on a substrate with a surface porosity of $P = 0\%$ and on a substrate with a surface porosity of $P = 6.2\%$ and humidity at application of paint of 9.9%. For coatings MA-115 on a substrate with $P = 6.7\%$ and humidity at the time of application of paint $W = 10.2\%$, peeling of the coating after 2 months of moistening is seen. Analysis of the experimental data (Table 3.6) shows that with increasing surface porosity of the substrate, a decrease in the stress intensity factor is observed up to a certain limit. Thus, with a load 25.39 N, the value of the stress intensity factor K_c for coating PF-115 on a substrate with a surface porosity $P = 0.13\%-1.2\%$ after 2 months of moistening is $K_c = 0.0376$MH/m$^{1.5}$, and on a substrate with a surface porosity, $P = 10.5\%$ it is 0.0256 MH/m$^{1.5}$; and for the MA-15 coating, the values of the stress intensity factor are, respectively, 0.0257 MH/m$^{1.5}$ and 0.02147 MH/m$^{1.5}$. Obviously, this is due to the fact that the pores on the surface of the substrate to some extent "extinguish" internal stresses and reduce the tendency to crack coatings. Increasing the substrate moisture at the time of application of the paint results in a more defective structure of the contact layer "coating-substrate" and a greater propensity to cracking. The stress intensity factor for PF-115 coatings at substrate moisture at the time of application of a paint composition equal to $W = 10.4\%$ after 2 months of

Table 3.6 Parameters crack education of protective-decorative coatings

Kind of colorful composition	**Substrate moisture,%**	**Porosity of the substrate,%**	**Load, *H***	**Coefficient of intensity of stresses, K_{1c}, MH/m^{1,5}**
1	2	3		5
Alkyd PF-115	0	0.13	25.39	0.0379
	0	0.9		0.0379
	0	10.5		0.0256
Alkyd PF-268	10.5	6.4		0.0404
	0	0.13		0.0554
	0	0.9		0.0554
	0	10.5		0.048
Oil MA-15	10.4	6.4		0.054
	0	0.33		0.0257
	0	6.4		0.02147
	9.9	6.4		peeling

wetting is $K_c = 0.404$ MH/m$^{1.5}$. Among other factors, the state of the painted surface of the facades of buildings is determined by the time of application of the paint. So, for example, if paint is applied in April–May, when the moisture of the substrate and the coating is high due to moisture migration from the side of the wall material, this can lead to premature failure of the coating. We evaluated the effect of substrate moisture on the properties of protective and decorative coatings, in particular, on their crack resistance. The analysis of the results (Table 3.7) shows that the substrate moisture at the time of application of paint has a significant effect on the crack resistance of the coatings. For example, when PVAC coating is applied to a dry surface, the appearance of cracks in the coating when Vickers indenter is introduced occurs after 20 cycles of freezing-thawing, with the critical value of the stress intensity factor being $K_{1c} = 0.09$ MH/ m$^{3/2}$. An increase in the initial moisture content of the substrate to $W = 1\%$ leads to a significant increase in the fracture toughness of the PVAC coating, with the appearance of cracks in the introduction of the Vickers indenter only after 25 cycles of freezing-thawing. Further increase in the substrate moisture during painting leads to the appearance of a more defective structure of the contact layer "coating-substrate" and a greater tendency of the coating to crack. Thus, an increase in substrate moisture to $W = 4\%$ led to the appearance of cracks with the introduction of the Vickers indenter after 10 cycles of freezing-thawing. The results of the analysis of fracture toughness indexes of coatings PF-115 indicate that after 15 cycles of freezing-thawing, coefficient of intensity of stresses in coatings on absolutely dry substrate is $K_1 = 0.057$ MH/m$^{3/2}$, whereas for substrate moisture $W = 4\%$ it is $K_1 = 0.026$ MH/m$^{3/2}$. Coatings PF-115 on a dry substrate are characterized by peeling after 15 cycles of freezing-thawing, while at substrate moisture content $W = 4\%$

Table 3.7 Parameters crack education of protective-decorative coatings depending on the initial moisture content of the substrate

Name of coating	**Moisture of the substrate, *W*,%**	**Impact type**	**Hardness *H*, mm²**	**The ratio of the half-length of the crack *C* to the size of the semi-diagonal of the imprint *a***	**Coefficient of intensity of stresses, K_{1c}, MH/m$^{1.5}$**
1	**2**	**3**	**4**	**5**	**6**
PVAC	0.0	After hardening	62	1	0.06
		5 freeze-thawing cycles	112	1	0.08
		10 freeze-thawing cycles	125	1	0.084
		20 freeze-thaw cycles	180	1.1	0.09[a]
PVAC	1.0	After hardening	32	1	0.018
		5 freeze-thawing cycles	80	1	0.022
		10 freeze-thawing cycles	102	1	0.08
		25 freeze-thawing cycles	112	1.08	0.09[a]
	4.0	After hardening	44	1	0.053
		5 freeze-thawing cycles	80	1	0.066
		10 freeze-thawing cycles	131	1.2	0.088[a]

PF-115	0.0	After hardening	48	1	0.006
		5 freeze-thawing cycles	87	1	0.033
		10 freeze-thawing cycles	111	1	0.05
		15 freeze-thawing cycles	142	1	0.057 peeling of the coating is observed
	1.0	After hardening	24	1	0.019
		5 freeze-thawing cycles	45	1	0.026
		15 freeze-thawing cycles	65	1	0.04
		20 freeze-thawing cycles	99	1	0.049
		30 freeze-thawing cycles	108	1	0.053
		35 freeze-thawing cycles	126	1	0.055 peeling of the coating is observed
PF-115	4,0	After hardening	18	1	0.017
		5 freeze-thawing cycles	37	1	0.024
		15 freeze-thawing cycles	44	1	0.026
		20 freeze-thawing cycles	63	1	0.038
		30 freeze-thawing cycles	86	1	0.045
		35 freeze-thawing cycles	97	1	0.05

[a]Vritical coefficient of intensity of stresses Thus, the value of the moisture content of the substrate, established in the regulatory and technical documentation, in particular, $W = 8\%$ for water paint, is not correct. There is a value for the optimum moisture content of the substrate for each particular coating in terms of its fracture toughness. It is necessary to conduct extensive research to create a databank on the influence of substrate moisture on the crack resistance of protective and decorative coatings. This will help in the future develop measures to create crack-resistant coatings.

peeling is not observed after 35 freeze-thaw cycles. The obtained results make it possible to assume that an increase in the moisture content of the substrate leads to a significant decrease in internal stresses in the coating, which is apparently due to a decrease in the adhesion strength due to the presence of moisture on the surface of the substrate. However, for different types of coatings, optimum moisture content of the substrate is characteristic. Increasing the optimum moisture content of the substrate leads to a decrease in the crack resistance of coatings.

References

[1] A.G. Evans, Structural Ceramics, 256, Moscow Publisher, Metallurgy, 1980.

[2] G.I. Gorchakov, L.P. Orentlicher, V.I. Savin, et al., Composition, Structure and Properties of Cement Concretes, Moscow Stroyizdat (1976) 144.

[3] V.I. Loganina, N.I. Makridin, L.V. Makarova, V.N. Karpov, Evaluation of the crack formation of coatings with the help of the acoustic emission method, Izvestiya Vuzov. Construction (6) (2003) 35–38.

[4] V.I. Loganina, L.V. Makarova, To a technique of an assessment of crack resistance of protective and decorative coatings, Plasts (4) (2003) 43–44.

[5] V.I. Loganina, L.V. Makarova, Evaluation of the influence of substrate quality on the crack resistance of protective and decorative coatings, Industrial coloring, 2005. N1, pp. 53–56.

[6] V.I. Loganina, L.V. Makarova, Influence of the coating structure and surface preparation on the crack resistance of paint coatings, Corrosion, materials, protection, 2005, No. 6, pp. 34–36.

[7] V.I. Loganina, L.V. Makarova, Evaluation of the effect of porosity of the substrate on the crack resistance of protective and decorative coatings. Izvestiya Tulskogo gosuniversiteta, Series "Building materials, structures and structures" (issue 4) (2003) 184–187.

[8] V.I. Loganina, N.I. Makridin, L.V. Makarova, O.V. Karpova, Evaluation of workability of protective and decorative coatings of cement-concrete concretes, Indus. Paint. (6) (2003) 29–31.

[9] S.I. Koryagin, S.V. Builov, Crack resistance assessment of a reinforced polymer coating applied to a metal surface 62 (10) (1996) 638–640.

[10] S.A. Nikulin, V.G. Khanzhin, A.B. Rozhnov, Analysis of crack resistance and quality of thin coatings by acoustic emission. 8th International Conference of the Slovenian-Society-for-Non-Destructive-Testing on the Application of Contemporary Non-Destructive Testing in Engineering. Portoroz, SLOVENIA. 2005 8th International Conference of the Slovenian Society for Non-Destructive Testing, Conference Proceedings: Application of contemporary non-destructive testing in engineering, pp. 309–316.

[11] V.I. Loganina, L.V. Makarova, S.N. Kislitsyna, Assessment of an Aging of Protective Decorative Coating,Contemporary Eng. Sci. 7 (36) (2014) 1961–1965. HIKARI Ltd, www.m-hikari.com https://doi.org/10.12988/ces.2014.411213.

Forecasting the durability of coatings

4

4.1 Regularities of aging of protective and decorative coatings of cement concretes

Moisture acts on the paint and varnish coating of the exterior walls of buildings during operation. The effect of moisture in the form of summer rains is relatively short-lived, and rainwater, penetrating in the coating, quickly evaporates. Strong autumn rains can wash out water-soluble compounds and cause mechanical damage to the coatings. The effects of dew are also dangerous, since the moisture condenses in the pores of the coatings, which leads to swelling of the coating and the appearance of whiteness and bubbles on its surface. Subsequent freezing can lead to the formation of cracks and peeling of coatings. As a result of the relative humidity of air, moisture sorption takes place, the value of which is determined by the relative humidity of the air and the properties of the coatings.

Water is an aggressive media and causes, over time, changes in the structure of the polymer, which are accompanied by the destruction of the chemical bonds of the main polymer chains. As a result of physical processes, as the moisture is absorbed by the coating a change in the molecular structure occurs along with a disruption of the pigment-film relationship, which leads to the initial stage of the destruction of the coatings, resulting in a change in color (the appearance of whiteness) and a loss of shine. According to Refs. [1–4], moisture penetrates primarily a dense layer along the boundaries of structural formations, inside the supramolecular structures. Penetration of moisture into the structure leads to an increase in volume and the appearance of larger supramolecular formations that cause a significant loosening of structures, their destruction and ultimately destruction.

Chemical changes in coatings under the action of moisture are due to the processes of hydrolysis and photohydrolysis. Hydrolysis accelerates with increasing air humidity and temperature [3].

In the process of moistening at the initial stage swelling is observed in coatings. The change in the mass of coatings occurs according to the law of a quadratic parabola,

$$\Delta m = kt^2 \tag{4.1}$$

where Δm is the change in mass of the coating, t is the time and k is a constant.

Increasing the Durability of Paint and Varnish Coatings in Building Products and Construction.
DOI: https://doi.org/10.1016/B978-0-12-817046-5.00004-8

At the first stage of moistening an increase in the shine of KO-168 coatings is observed, which is apparently due to the plasticizing action of moisture that facilitates of relaxation processes. Spontaneous leveling of the surface microrelief also occurs. These data show that at the first moment of moistening in the silicone-organic coating of KO-168 swelling processes prevail. The process speed is controlled by diffusion. After 18–20 hours of moistening the change in the mass of the coatings follows other regularities, obviously due to the simultaneous occurrence of the processes of swelling and destruction. After 450 hours of humidification destructive processes predominate, expressed by a decrease in the shine of the coatings. It is established that the rate of destruction depends on the amount of sorbed moisture (Fig. 4.1).

For PVAC coatings a different fracture mechanism is observed. In the first moment of moistening the processes of swelling and structure formation take place simultaneously. This is evidenced by the data on the change in mass and hardness of the coatings after 10 hours of moistening. The increase in the hardness of coatings is 6%. Later, the hardness of coatings is reduced, due to the processes of destruction.

For VD-AK-111 coatings at the initial stage of moistening swelling is observed. After 300 hours of moistening, the water absorption of the coatings is 16.6%. After 600 hours of humidification a decrease in the mass of the coatings is observed,

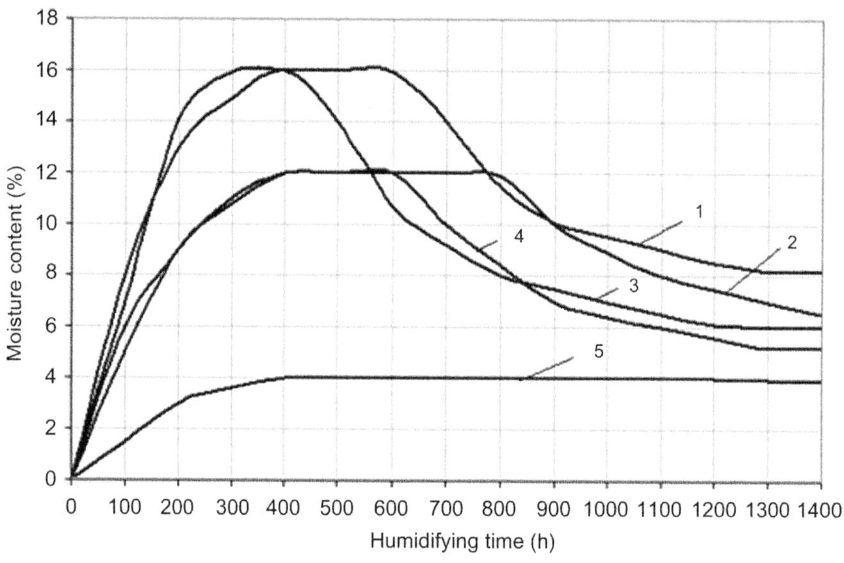

Figure 4.1 Kinetics of moisture of coatings.
1, 3—VD-AK-111 coating
2, 4—PVAC coating
5—KO-168 coverage
1, 2, 3—Temperature 20°C
3, 4—Temperature 60°C

which is obviously related to the hydrolysis of the binder. The increase in temperature promotes more rapid destruction of coatings. A decrease in the mass of coatings AK-111 was observed after 500 hours of moistening at a temperature of 40°C, and in PVAC coatings after 650 hours of moistening (Fig. 4.2).

At the first moment of humidification of coatings a sharp increase in internal stresses is characteristic, then a slow relaxation followed by stabilization (Fig. 4.3). Thus, for example, in the initial period of moistening for polymer coatings (up to 18 days) an increase in internal stresses up to 0.19 MPa is observed, and after 18 days of humidification their stabilization at the level of 0.062 MPa occurs.

For PVAC coatings a temporary factor is also characteristic; an increase in internal stresses (in the first 4–5 days of moistening) is observed, followed by slow relaxation and after 26 days of moistening their stabilization at the level of 0.08 MPa is seen.

Humidification causes a sharp change in the modulus of elasticity of coatings. For example, the modulus of elasticity of PVAC coatings in the initial state was 1.31×10^2 MPa, and after humidifying for 25 days it was 0.52×10^2 MPa. The decrease in the modulus of elasticity at moistening was of the same order as the decrease in hardness. A sharp decrease in the modulus of elasticity during moistening reduces the risk of cracking during deformation of the wall structure.

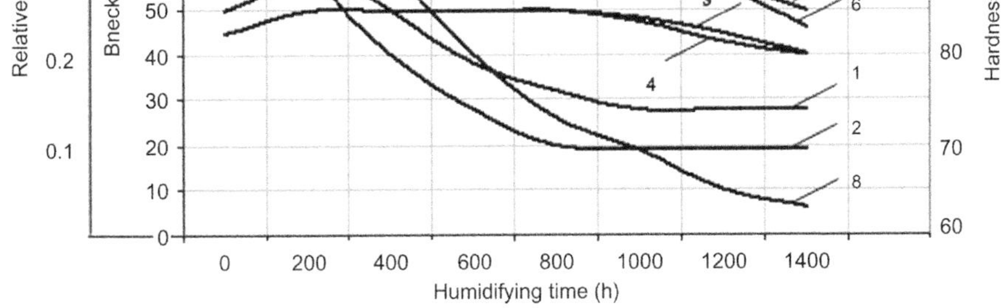

Figure 4.2 Changes in the properties of coatings during moistening.
1, 2—The relative hardness of VD-AK-111 coating
3, 4—Shine of PVAC coatings
5, 6—Shine of WD-AK-111 coatings
7—Shine of KO-168 coatings
8—Hardness of KO-168 coatings
1, 3, 5—Temperature 20°C
2, 4, 6—Temperature 60°C

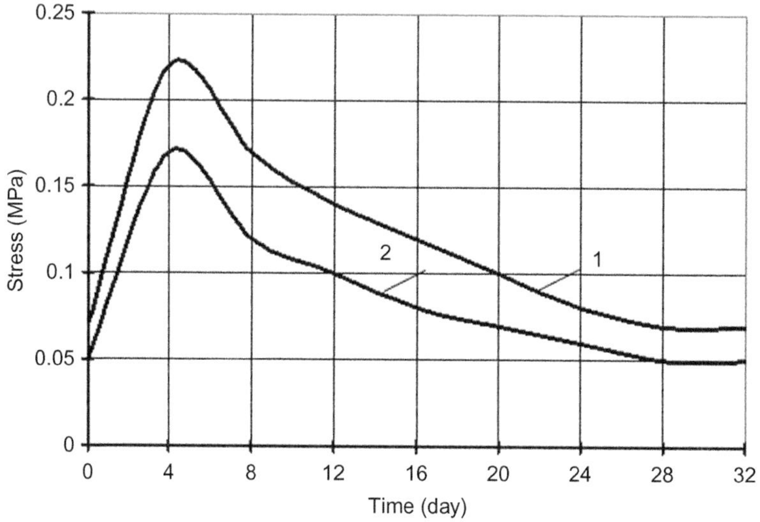

Figure 4.3 Changes in internal stresses during the moistening process.
1—PVAC coating
2—Polymer-calcareous coating

When moistening the elastic part of the deformation decreases. For example, the fraction of the elastic component of PVAC coatings before wetting was 0.19, and after wetting was 0.13; for the lime coatings, respectively, 0.104 and 0.074; for the polymer-lime coating 0.08 and 0.05, respectively.

The duration of preservation of cohesive strength of coatings during operation is also determined by the resistance to periodic effects of environmental factors: moistening-drying, freezing-thawing, etc. With this in mind, the influence of moisture on the change in the duration of preservation of cohesive strength was assessed. Samples of the polymer-lime, PVAC coatings and coatings from a dry finishing mixture after 3 days of moistening were dried for 2 weeks at an air temperature of 18°C–20°C and a relative humidity of 60%–70%, after which they were tested for tension. Table. 4.1 shows the data on the changes in the physicomechanical properties of coatings.

The results of experimental research show that drying of the PVAC coatings after 3 days of humidifying promotes further hardening of structure of the coating. Thus, the initial value of cohesive strength, $R_p = 2.22$ MPa, and at the subsequent drying after humidifying, $R_p = 3.56$ MPa. Similar results are also seen for the polymer-limy coatings and for the coatings on the basis of dry finishing mixture. Thus, for coatings on the basis of dry finishing mixture, the initial value of cohesive strength is $R_p = 1.66$ MPa, and after drying $R_p = 2.32$ MPa.

As the results show, after 3 days of humidifying the share of destructive processes in the mechanism of aging of coatings on a basis polymer-mineral binding is insignificant as the subsequent drying not only restores initial cohesive strength, but also promotes its increase.

With increasing the duration of humidifying there is an incomplete restoration of properties of coatings, that is, there are irreversible phenomena. This is evidenced by data on changes in the strength of adhesion for the PVAC as well as polymer-limy coatings (Table 4.2).

Thus, the initial adhesive strength for the PVAC coatings (before tests) is $R_{ad} = 2.2$ MPa, and after 10 days, there is an incomplete recovery of the adhesive

Table 4.1 Physicomechanical properties of coatings

The name of parameters	Before tests	After humidifying and subsequent drying
PVAC coating		
Cohesive strength (MPa)	2.22	3.56
The modulus of elasticity, $E \times 10^2$ (MPa)	1.31	1.34
Elongation ε (%)	1.76	2.87
Plastic deformation, $\varepsilon_{\text{пл}}$ (%)	0.1	0.13
The polymer-calcareous coating		
Cohesive strength (MPa)	1.87	3.17
The modulus of elasticity, $E \times 10^2$ (MPa)	1.03	1.34
Elongation ε (%)	1.00	1.6
Plastic deformation, $\varepsilon_{\text{пл}}$ (%)	0.5	0.26
The coating on the basis of dry finishing mixture		
Cohesive strength (MPa)	1.66	2.32
The modulus of elasticity, $E \times 10^2$ (MPa)	1.31	1.51
Elongation ε (%)	1.16	1.26
Plastic deformation, $\varepsilon_{\text{пл}}$ (%)	0.10	0.2

Table 4.2 Restoration of strength of adhesion of coverings after humidifying and subsequent drying

Kind of a coating	Adhesive strength (MPa)						
	Age of tests (days)						
	0	**10**	**20**	**30**	**50**	**90**	**120**
PVAC	2.42	2.15	1.98	1.75	1.6	1.4	1.23
		2.3[a]	2.2	1.93	1.86	1.79	1.6
The polymer-calcareous	2.2	1.9	1.72	1.5	1.45	1.2	1.09
		2.1	1.9	1.62	1.56	1.4	1.15

[a]Above the line feature values of adhesive strength of coatings after humidifying; below the line after drying of coatings.

strength, which is as much as 95.45% of the original after drying. Moreover, with increasing duration of humidifying, the proportion of irreversible processes increases, that is, the adhesion strength of the coating after drying is less and less compared to the initial value. Thus, after 120 days of humidifying, the adhesive strength of the PVAC covering becomes as much as 66% (after drying) of the initial value. Similar results are seen for the polymer-limy coating.

Consider the aging process of the coating when moistened by changing the operational properties indicators. We denote by T_o the initial value of the hardness of the coating and by T_t the hardness at the time t. The rate of change in hardness will be

$$\frac{dT}{dt} = k(T_o - T_t) \tag{4.2}$$

where k is the constant of rate of change of hardness.

Solving Eq. (4.2), we obtain

$$T_t = T_o[1 - \exp(-kt)] \tag{4.3}$$

or

$$t = \frac{1}{k}\ln\frac{T_o}{T_o - T_t}$$

However, the results of the studies show that the constant of change in hardness is not a constant. This indicates that the process of changing hardness is a reflection of the course of several parallel processes. It is established that the rate constant varies according to the law (Fig. 4.4):

$$k = 1/(A + Bt) \tag{4.4}$$

After substituting Eq. (4.4) into Eq. (4.3) and mathematical transformation, we obtain

$$t = A\frac{\ln\frac{T_o}{T_o - T_t}}{1 - B\ln\frac{T_o}{T_o - T_t}} \tag{4.5}$$

Eq. (4.5) allows us to determine the aging time of the coating when moistened to a specified hardness T_t. The calculation shows that the values of A and B for PVAC coatings are -4 and -1.1856 and for the lime -6 and -2.1093, respectively.

The rate constant of the structure formation can be represented by the equation

$$\frac{dC}{d\tau} = k_1 C \tag{4.6}$$

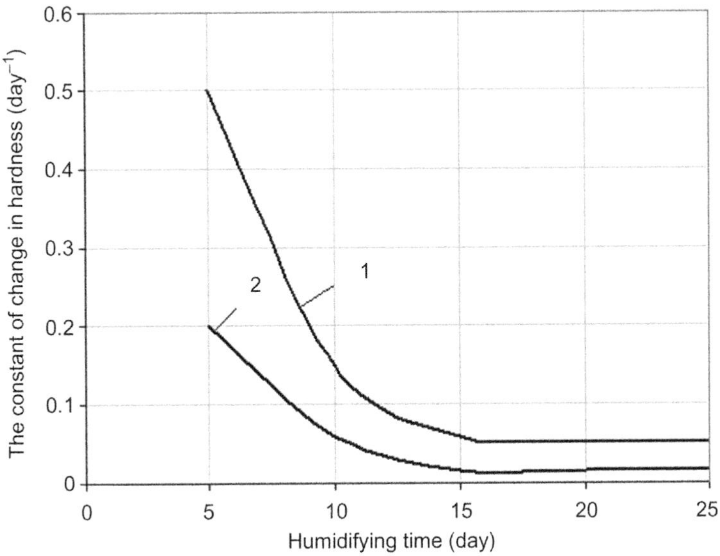

Figure 4.4 Dependence of the rate constant of change of the hardness of coatings on the duration of moistening.
1—Lime
2—PVAC

where C is the concentration of substances, $k_1$is the constant rate of structure formation and$\tau$is the time of the course of structure-forming processes during moistening.

The change in the concentration of reactants is proportional to changes in the properties of the coatings. Eq. (4.6) can be represented by changing the indicators characterizing the resistance of the coatings. If we use the hardness T as such a parameter, then Eq. (4.6) will be

$$\frac{dT}{d\tau_1} = k_1 T \tag{4.7}$$

Taking into account the course of destructive processes, the change in hardness in time can be described by

$$T = T_1 \exp(-k_1 t_1) - T_2 \exp(-k_2 t_2) \tag{4.8}$$

where k_2 is the rate constant of destructive processes and t_2 is the time of the course of destructive processes.

To determine k_1, k_2, T_1, T_2, we use the following arguments. For $t_1 < t_2$, the rate of destructive processes is small, and thus Eq. (4.8) can be written as

$$T = T_1 \exp(-k_1 t_1) \tag{4.9}$$

The graphical solution of Eq. (4.9) made it possible to obtain the value of the rate constant for the structure formation in the process of wetting the PVAC coating, which is equal to $k_1 = 1.97 \times 10^{-7}\ s^{-1}$. The numerical values of the rate constants for the destructive processes were found by graphical solution of Eq. (4.9) after substitution of numerical values k_1 and T_1. For PVAC coverage $k_2 = 4.0531 \times 10^{-4}\ s^{-1}$. The higher numerical value of the rate constant k_2 of the destructive processes the higher the predominance of destructive processes in the aging mechanism of the PVAC coating.

To determine the effective activation energy for structure and destruction, the rate of change in the mass of coatings during the moistening process at different temperatures was also determined. It was found that the effective activation energy for destruction after 1200 hours of moistening the PVAC coating is $U = 0.976$ kJ/mol, and for coatings based on water-dispersion paint WD-AK-111 it is 7.158 kJ/mol. The total effective activation energy of aging of the coating after 1200 hours of moistening with a 10% loss of brightness is $U = 5.159$ kJ/mol for PVAC, and for coating based on AK-111 paint, 8.39 kJ/mol.

The experimental and calculated data allow us to understand in more detail the mechanism of aging of coatings and to establish the factors that regulate the rate of structure-forming and destructive processes, which will improve the life of coatings.

This research shows the complexity of the mechanism of aging of coatings of cement concretes. Let us consider the processes that take place on the surface of the coatings during moistening. When moistening polymeric coatings the aging process can be considered as a heterogeneous process in which the diffusion transfer of a substance is accompanied by a chemical interaction.

Diffusion of the substance (moisture) occurs from phase 1 (external environment) to phase 11 (coating), in which the chemical reaction of the interaction of moisture and oxygen with the polymer binder takes place).

Suppose that in phase 1 the moisture concentration is $c = c_o$. Accounting for the balance of the substance at diffusion leads to the equation

$$Dd^2c/dx^2 = dc/dt \tag{4.10}$$

The hydrolysis reaction of the polymer binder has the first order. Then the equation of material balance has the form

$$Dd^2c/dx^2 = dc/dt + kc \tag{4.11}$$

When a stationary state occurs $dc/dt = 0$. To solve Eq. (4.11), it is necessary to find two of its linearly independent particular solutions. Let be $y = \exp(rx)$, then

$$y = r^2\exp(rx) \tag{4.12}$$

$$Dr^2\exp(rx) - k\exp(rx) = 0 \tag{4.13}$$

$$\exp(rx)[Dr^2 - k] = 0 \tag{4.14}$$

As

$$\exp(rx) = 0$$

then

$$Dr^2 = k, r = \pm\sqrt{k/D} \tag{4.15}$$

The general solution of Eq. (4.11) has the form

$$c(x) = A\exp(-\sqrt{k/D})x + B\exp(\sqrt{k/D}x) \tag{4.16}$$

When $x = 0$ $c = c_o = A + B$. When $x = x_{\max}$ $c = 0$. This is possible for $B = 0$, therefore

$$c = c_o\exp(-\sqrt{k/D}x) \tag{4.17}$$

The process speed at the surface, that is, in the points, where $x = 0$.

$$v = -D\frac{dc}{dx} \tag{4.18}$$

The differentiation of Eq. (4.17) leads to the form

$$dc/dx = -\sqrt{k/D}c_o\exp(-\sqrt{k/D}x \tag{4.19}$$

when

$$x = 0 \quad dc/dx = -c_o\sqrt{k/D} \tag{4.20}$$

Substituting Eq. (4.20) into Eq. (4.18), we obtain

$$y = -c_o\sqrt{k/D} \tag{4.21}$$

Since $D = D_o\exp(-U/RT)$, after substituting the expressions for the diffusion coefficient and the rate constant into Eq. (4.21), we obtain

$$k = k_o\exp(-U/RT) \tag{4.22}$$

$$v = c_o\sqrt{D_o}k_o\exp[(-U-U)/2RT] = c_o\sqrt{D_o}k_o\exp(U_{эфф}/RT) \tag{4.23}$$

Depending on the ratio D and k, the process can take place in the diffusion, kinetic or transition regions. If we depict the dependence of the process velocity on the temperature in coordinates lg $v - 1/T$, we can get a clear idea of the areas of the process. The section, parallel to the abscissa axis expresses the internal diffusion region. The section of the curve AB corresponds to the flow in the internal kinetic region, from point A to B transition (Fig. 4.5).

In Fig. 4.6 the dependences of the rate of change in the shine of organosilicon coatings with the loss of shine of 20% on the ambient temperature are given. The form of the graphical dependence shows that the aging process of coatings occurs in the kinetic region. Additional proof of this is the numerical values for the activation energy of diffusion of moisture. Thus, for example, the activation energy of moisture diffusion U for a PVAC coating is $U = 5.7$ kJ/mol, modified coating of PVAC with addition of GKZh-11 −6.5 kJ/mol, for organosilicon KO-168 −9.1 kJ/mol. According to numerical values, the activation energy of diffusion turns out to be much less than the activation energy of the chemical interaction. This also indicates that the aging process of coatings during moistening takes place not in the diffusion but in the kinetic region. The rate of aging can be determined by the rate of the chemical reaction.

Proceeding from this, the properties of the coatings during moistening (adhesion strength, shine, hardness of coatings, etc.), in which various chemical processes predominate, will be described by an equation of the form

$$y = y_o \exp(-\alpha t) \tag{4.24}$$

where y is the value of the coating property after moistening, t is time of moistening, α is a coefficient characterizing the rate of change in the property and y_o is initial value of the coating property.

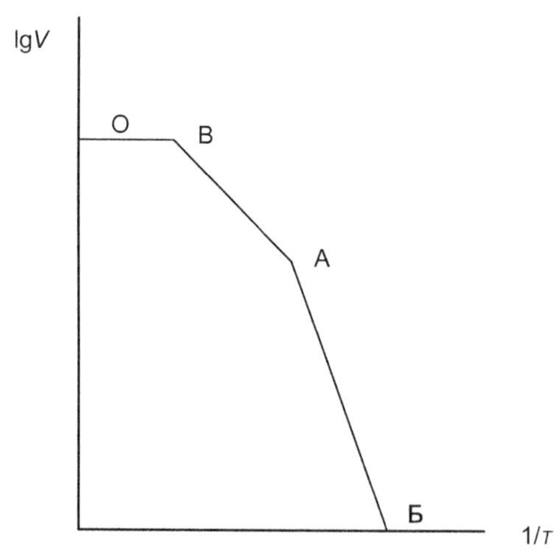

Figure 4.5 Dependence of the reaction rate on temperature.

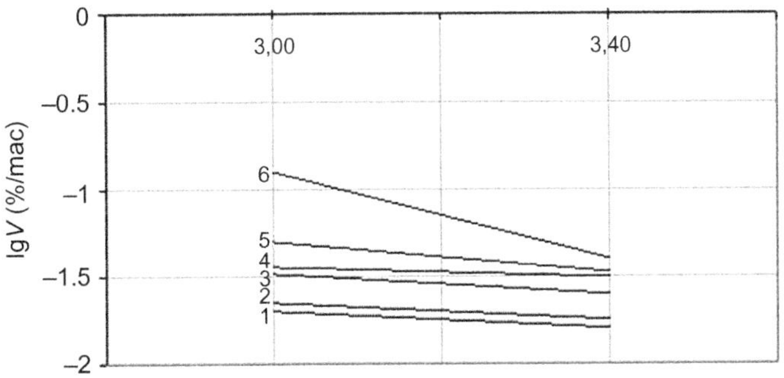

Figure 4.6 Dependence of the shine loss rate of coating KO-168 on the magnitude of the inverse absolute temperature with a loss of shine of 20%.
1—Coating on a glass substrate, $\phi = 60\%$
2—Coating on a glass substrate, $\phi = 100\%$
3—Coating on a cement substrate $P = 20.3\%$, $\phi = 60\%$
4—Coating on a cement substrate, $\phi = 100\%$
5—Coating on a cement substrate, $P = 26.6\%$, $\phi = 60\%$
6—Coating on a cement substrate, $\phi = 100\%$

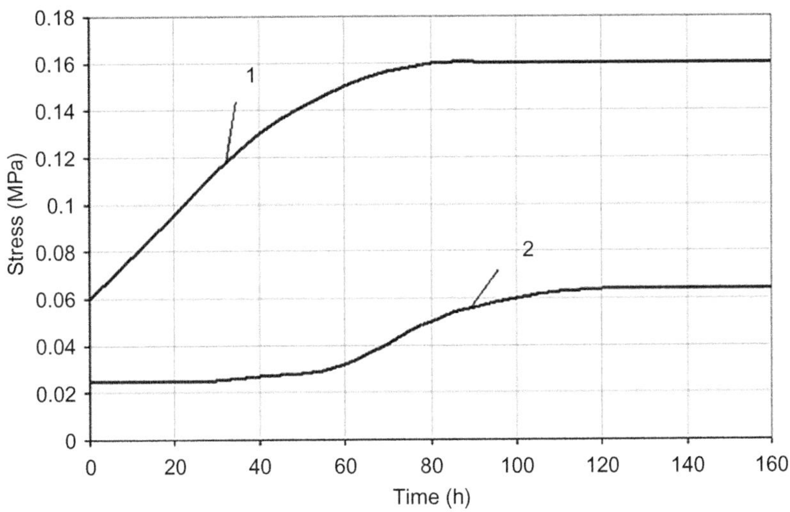

Figure 4.7 Changes of internal stresses in the process of heat aging.
1—PVAC coating
2—Polymer-calcareous coating

This makes it possible, using the laws of chemical kinetics, to determine the speed, and consequently, the aging time.

At studying the regularities of thermal aging of coatings, internal stresses were measured. The coating samples after they were cured, were in a drying cabinet at a temperature of $t = +50°C$. Analysis of the experimental data (Fig. 4.7) indicates that at aging of coatings under the influence of positive temperatures, internal stresses are increase. Stabilization of internal stresses is observed in polymer coatings after 80 hours at 0.06 MPa level and in PVAC coatings after 90 hours at the level of 0.162 MPa.

The physical and mechanical properties of coatings change during thermal aging. In the initial period of thermal aging, increase in the hardness of polymer-mineral free films is observed, evidently due to the processes of structure formation. In this case, the rigidity of the films increases, and an increase in the modulus of elasticity is observed. The increase in the hardness of the coating, and, consequently, of the internal stresses also indicates that further coatings are subjected to curing (Table 4.3).

These data are in good agreement with the results obtained when testing free films on a tensile machine. Analysis of the experimental data (Table 4.4) indicates

Table 4.3 Changes in hardness of coatings during thermal aging

Kind of a coating	**Hardness (MPa)**			
	Duration of heat aging (h)			
	0	**50**	**100**	**150**
PVAC	41.14	66.28	61.11	49.46
Calcareous	35.47	53.44	49.28	47.71

Table 4.4 Changes in physical and mechanical properties of coatings in the process of thermal aging

The name of parameters	**Duration of heat aging (h)**			
	0	**50**	**100**	**150**
PVAC coating				
Cohesive strength (MPa)	2.22	5.49	4.62	4.1
The modulus of elasticity, $E \times 10^2$ (MPa)	1.31	2.92	2.54	1.94
Elongation ε (%)	1.76	2.9	2.65	2.4
Plastic deformation, $\varepsilon_{пл}$ (%)	0.1	1.05	1.1	1.05
The polymer-calcareous coating				
Cohesive strength (MPa)	1.87	–	2.11	2.04
The modulus of elasticity, $E \times 10^2$ (MPa)	1.03	–	1.8	1.64
Elongation ε (%)	1.0	–	1.9	2.0
Plastic deformation, $\varepsilon_{пл}$ (%)	0.5	–	0.7	1.33

that in the initial state (before the onset of aging), the presence of elastic and plastic deformations is typical for polymer and PVAC films. Thus, the value of plastic deformations during stretching of PVAC films is $\varepsilon = 0.1\%$, and the total elongation is $\varepsilon = 1.76\%$.

During thermal aging an increase in cohesive strength is observed, which is caused by the processes of structure formation. But after 50 hours of testing, destruction processes prevail. After 50 hours of testing the relative deformations are $\varepsilon = 2.9\%$, and the residual strain $\varepsilon = 1.05\%$. With an increase in the duration of thermal aging of PVAC films the numerical values of the relative elongation decrease and the proportion of residual deformations in the total deformation of the films increases. Thus, after 50 hours of thermal aging, the proportion of permanent deformation is 35%, while after 150 hours it is 44%.

Thermal aging causes a decrease in the elastic modulus of the coatings over time. After 150 hours of thermal aging, the elastic modulus of PVAC coatings is $E = 1.94 \times 10^2$ MPa.

Results of experimental studies show that polymer mineral coatings behave like brittle bodies. After thermal aging the curve of the dependence $R_p = f(\varepsilon)$ characterizes coatings as elastically fragile bodies with some plasticity. An increase in the relative elongation is observed, that is, there is a transition from brittle fracture to elastic-plastic.

In the process of thermal aging a change in the adhesion strength of coatings is observed. The graph of the dependence of the adhesion strength on the duration of thermal aging is shown in Fig. 4.8.

Analysis of the experimental data indicates that the adhesion strength of coatings under the action of temperature decreases exponentially ,which agrees with the experimental data of other authors [1–3].

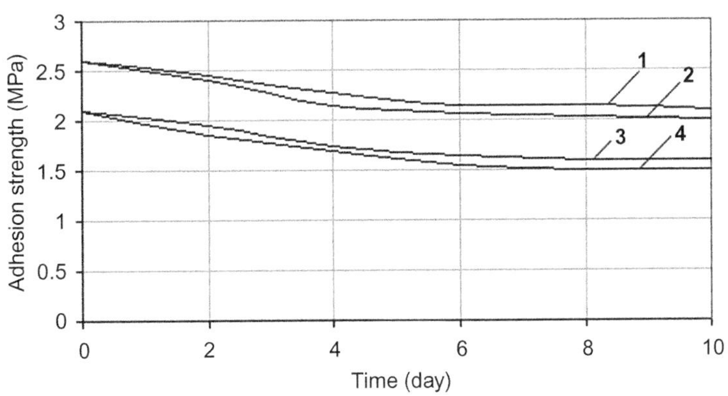

Figure 4.8 Adhesion strength of the coating as a function of temperature:
1, 2—PVAC coating
3, 4—KO-168 coating
1, 3—Temperature 40°C
3, 4—Temperature 60°C

The aging rate constant can be represented by the equation

$$-\frac{dC}{dt} = kC \tag{4.25}$$

where C is the concentration of reactants.

The change in the concentration of reactants promotes proportional change in the properties of the coatings. Therefore, Eq. (4.25) can be represented in terms of the change in the indices characterizing resistance of coatings. As such a parameter, we propose the use a generalized assessment of decorative and protective $A3$ properties [12]:

$$-\frac{dA3}{dt} = kA3 \tag{4.26}$$

The state of "failure" of the coating is characterized by the value $A3 = 0.44$. Therefore, Eq. (4.26) after integration has the form

$$\int_{1.0}^{0.44} dA3/A3 = kA3 \tag{4.27}$$

Thc time to reach the value of $A3 = 0.44$ for different temperatures can be determined by the relation

$$k_1 t_1 = k_2 t_2 = k_i t_i = k \int_{t_1}^{t_2} dt \tag{4.28}$$

Then

$$k = 0.82/\tau(T) \tag{4.29}$$

Substituting Eq. (4.29) into Eq. (4.27), we obtain

$$k \int_{t_1}^{t_2} dt = 0.82 \int_{t_1}^{t_2} dt/\tau(T) = 0.82 \ln\left|\tau(T)\right|_{t_1}^{t_2} \tag{4.30}$$

Eq. (4.30) characterizes the temperature dependence of the time for the protective walls to reach the "failure" state.

The destruction of coatings under the influence of sunlight in the initial stage of aging is due to photooxidative destruction. During the oxidation of coatings, liquid and gaseous degradation products are formed, which contribute to the process of structure formation. The change in the morphology of the surface of coatings under the influence of light radiation confirms the course of the processes of structure formation during aging of coatings.

Formed in the process of curing coatings, supramolecular structures with aging show more clearly. In this case, growth of structural elements, their aggregation and the transition of simple forms of supramolecular structures to more complex ones are observed. With the aging of pigmented coatings, the structure formation processes lead to the aggregation of pigment particles and the occurrence of significant stresses around large aggregates [3,4]. The most probable are the processes of photooxidation at the boundaries of structural formations. The growth of internal stresses leads to an increase in the number of discontinuities in macromolecules and an increase in radical decomposition reactions, which contributes to the destruction of coatings.

According to the data of Refs. [5,6], the inner layers of coatings in comparison with the surface layers have supramolecular structures of a larger size. In the process of aging and the destruction of the surface layers more brittle inner layers are exposed. When the surface layer is destroyed, the number of aggregates protruding above the surface of the coating increases. The gradual destruction of the surface layer as well as the replacement of smaller structural formations by larger ones (after aging) indicate that under the influence of UV irradiation there is a loosening of the structure and an increase in the number of pores, cracks and defects. This leads to an increase in the surface roughness of the films and, consequently, the surface area of the coating. Under UV irradiation, during the initial aging period, intensive destruction of the coatings is observed, then the rate of destruction gradually decreases.

For AK-111 coatings after 120 hours of aging, the decrease in mass m is 55%, for coatings PVAC- 45%. Analysis of the data in Fig. 4.9 shows that the dependence of the loss of shine and the change in the mass of coatings on the duration of aging can be described by an equation of the form

$$y = ax^b \tag{4.31}$$

This indicates that a change in the shine (characterizing the ehe destruction of coatings) is associated with the loss of mass in the surface layer 1 μm thick. The method of least squares generated calculations of the constants a and c in Eq. (4.31). For AK-111 coatings the equation is

$$m = 27.12\tau^{0.1666}$$

For PVAC coating

$$m = 4.66\tau^{0.54}$$

Analysis of the data (Fig. 4.10) shows that under the influence of UV irradiation coatings exhibit an increase in internal stresses. However, after t hours of irradiation, relaxation of internal stresses is observed. Thus, for example, for PVAC coatings, relaxation of internal stresses is observed after 40 hours of UV irradiation, evidently due to the development of plastic deformations.

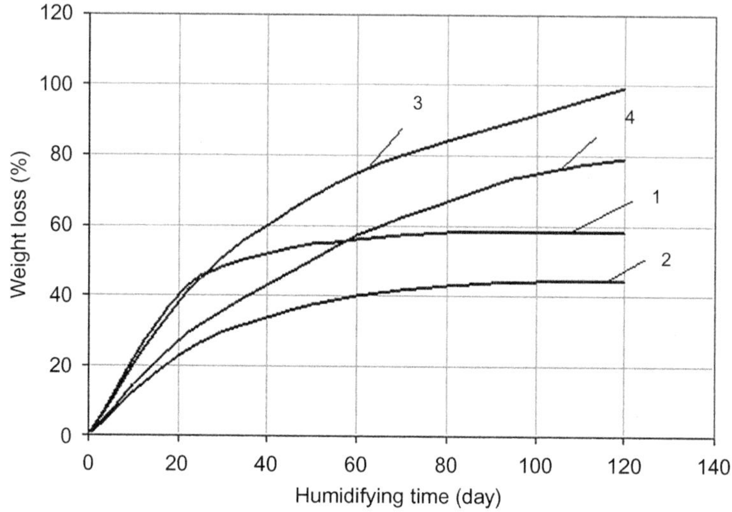

Figure 4.9 Loss of mass and shine of coatings with UV irradiation.
1, 3—WD-AK-111coating
2, 4—PVAC coating
1, 2—Weight loss
3, 4—Loss of shine

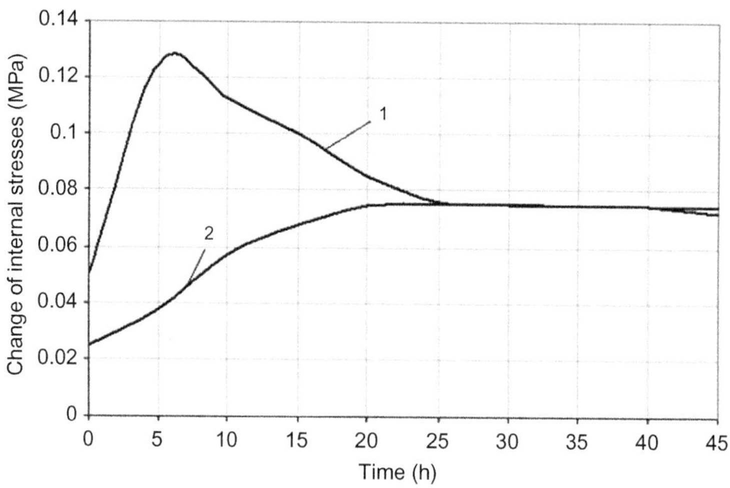

Figure 4.10 Changes in internal stresses during UV irradiation.
1—PVAC coating
2—Polymer-line coating

The results of tests on a tearing machine IR 5057-50 show (Table 4.5) that UV irradiation causes an initial increase in cohesive strength with a subsequent slight decrease, apparently due to the slow rate of destructive processes on this the stage of aging. Thus, PVAC coating after 50 hours of UV irradiation is characterized by $R_p = 4.62$ MPa and a relative strain $\varepsilon = 2.7\%$, with the initial value of $R_p = 2.22$ MPa and $\varepsilon = 1.76\%$.

This is obviously explained by the ongoing processes of structure formation at the initial stage of aging. Further reduction of cohesive strength indicates the beginning of the processes of destruction in coatings. Thus, after 150 hours of UV irradiation, the strength of PVAC coatings begins to decrease from 4.62 MPa (50 hours of UV irradiation) to 4.36 MPa and plastic deformations decrease from 2.7% to 2.3%.

The combined effects of climatic factors (UV irradiation, moisture, temperature) change the nature of the aging process. Analysis of the experimental data (Fig. 4.11A) shows that during the first period of aging the swelling processes predominate. They are accompanied by a slight decrease in shine and a strong change in the adhesion strength of organosilicon KO-168 coatings, for example.

After 10 days of testing, the change in the adhesion strength of coatings is 0.023 MPa/day and shine 0.1%/day. In the period from 15 to 30 test cycles, swelling and destruction of coatings predominate, and after 30 cycles, the processes of destruction, swelling, and chemical hardening of coatings do. For PVAC coating a different failure mechanism is observed (Fig. 4.11B). In the beginning (up to 5 cycles), the processes of structure formation and destruction predominate, as can be seen by the data on the increase in the hardness of the coatings and in the reduction of the strength of adhesion. After 5 test cycles the processes of destruction and

Table 4.5 Changes in physical and mechanical properties of coatings in the process of aging under the influence of UV irradiation

The name of parameters	**Duration of exposure (h)**			
	0	**50**	**100**	**150**
PVAC coating				
Cohesive strength (MPa)	2.22	4.62	4.4	4.36
The modulus of elasticity, $E \times 10^2$ (MPa)	1.31	–	2.89	1.37
Elongation ε (%)	1.76	2.7	2.45	2.3
Plastic deformation, $\varepsilon_{пл}$ (%)	0.1	1.1	1.05	1.0
The polymer-calcareous coating				
Cohesive strength (MPa)	1.87	1.86	2.1	1.68
The modulus of elasticity, $E \times 10^2$ (MPa)	1.03	2.08	–	1.12
Elongation ε (%)	1.01	2.3	2.1	2.2
Plastic deformation, $\varepsilon_{пл}$ (%)	0.5	0.4	0.45	1.6

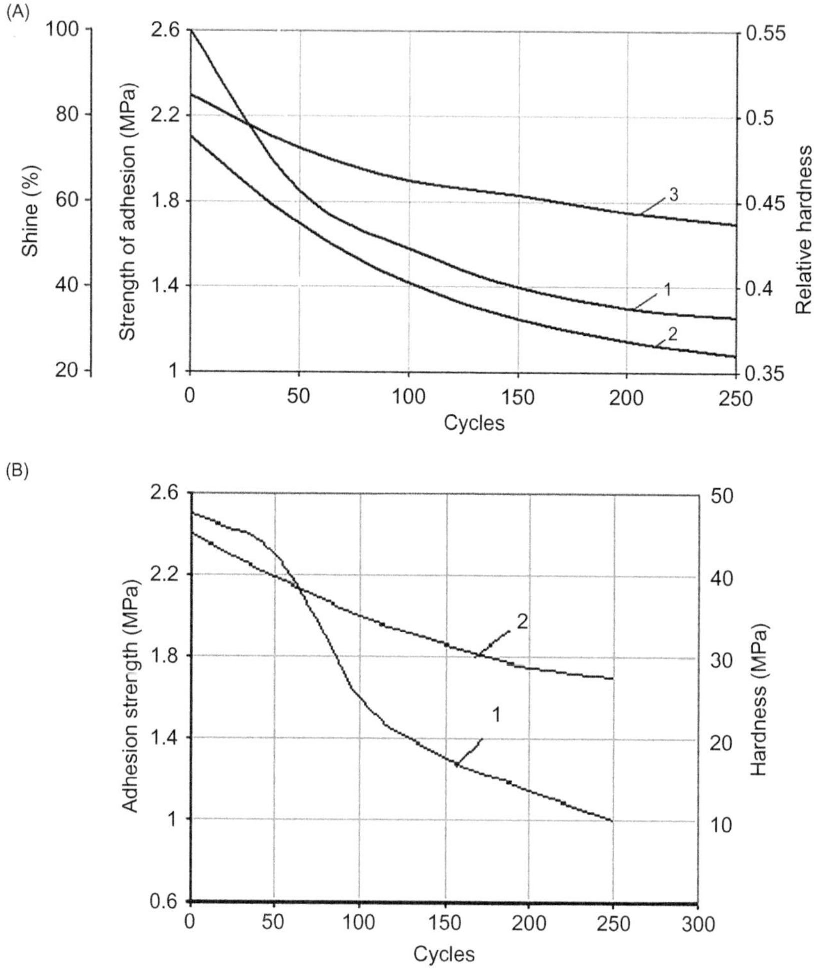

Figure 4.11 (A) Changes in the properties of organosilicon in the KO-168 coating during aging.
1—Hardness of the coating
2—Adhesion strength of coating
3—Shine of coating
(B) Change in the properties of PVAC cement coating during aging.
1—Hardness of the coating
2—Adhesion strength of the coating

swelling prevail, and after 35 cycles of testing, swelling, structure formation and destruction proceed simultaneously, as can be seen by data that shows a slight change in hardness and a very strong change in bond strength.

When assessing the aging kinetics of cement concrete coatings it was found that the change in the protective properties of coatings in the initial period of operation and in the forced mode tests is insignificant, that is, the rate of damage growth is zero. After 7 cycles of forced tests an active stage of accumulation of damage to the lime coating is observed. In full-scale tests, this time for calcareous coating was 90 days.

As already noted above, during the aging of protective and decorative coatings of the exterior walls of buildings, irreversible changes in their protective and decorative properties occur. The process of destruction of coatings is preceded by the stage of accumulation of damage of various scales, from the rupture of individual chemical bonds to the occurrence of submicro- and microscopic cracks before the main crack is formed [7,8]. The numerical values of the properties depend on the time, separating the present moment t from the moment t, t_f [0; t), and also from the properties of the coatings at time t. The kinetics of changing the properties of coatings on a given time interval is also determined by the prehistory of aging.

For a confirmation of this assumption the following experiment was done. Samples of protective-decorative coatings on a mortar's base were tested. Water-dispersion (WD-AK-111) and the polyvinyl-acetate-cement (PVAC) paint were used. The influence of thermal aging as well as humidifying in various sequence for properties of coverings such as relative hardness (the WD-AK-111 coating) and durability of cohesion with the substrate (PVAC coating) were studied. The samples were humidified during 800 hours and then exposed to thermal aging at the temperature of 60°C over 240 hours. Since there was a change of the factor impact sequence, the environmental parameters were kept stable.

The coefficient of additivity K_a was used to quantitatively estimate the influence of the factors [9–11]. It is determined by a ratio of changes in the properties for a given sequence of factors to the sum of the changes caused by each factor separately.

$$K_a(t, W) = \Delta f(t, W)/\{\Delta f(W) + \Delta f(t)\} \tag{4.32}$$

$$K_a(W, t) = \Delta f(W, t)/\{\Delta f(W) + \Delta f(t)\} \tag{4.33}$$

where $\Delta f(W), \Delta f(t)$ is the change properties of the coatings after moistening and thermal aging; respectively; t_2 is change properties of the coatings after moistening and subsequent aging; and $\Delta f(t, W)$ is change properties of the coatings after thermal aging and subsequent moistening.

After changing the coatings' properties Δf was defined as the difference between an initial parameter (before aging) and a parameter after aging. The results are shown in Table 4.6.

Table 4.6 Additivity coefficient at various sequences of climatic factors

Coating type	$K_a(W, t)$	$K_a(t, W)$
WD-AK-111	1	0.82
PVAC	0.85	0.72

The results of these studies show that in the process of moistening an exponential decline in the relative hardness of the WD-AK-111 coatings and decreasing adhesion of the PVAC coatings occurs.

Thermal aging of the WD-AK-111 coatings at a temperature of 60°C during 240 hours causes an increase in the relative hardness of the WD-AK-111 coatings from 0.45 to 0.48. In the PVAC coatings, decrease of adhesion to the substrate during thermal aging is observed.

PVAC coatings show a decrease in adhesion strength to the substrate during thermal aging. The influence of the climatic factors on the change in the properties of the coatings manifests in different ways. For PVAC coatings, if thermal aging follows moistening, then it causes a smaller change in the adhesion strength as compared to the sum of the individual impacts. When changing the order of influence of climatic factors, that is, moistening preceded by thermal aging over 240 hours, weakening of destructive influence of factors is observed. For WD-AK-111 coatings a slightly different dependence of the sequence of climatic factors is observed. If the thermal aging follows after moistening for a given degree of intensity of the factors, the total effect of the change in the relative hardness of the coating is equal to the additive effect of the climatic factors. If thermal aging precedes moistening, there is a weakening of the effect of the climatic factors.

The results of the calculations show that the value of the additivity coefficient changes depending on the sequence of the climatic factors and the type of covering (Table 4.6). Thus, $K_a(W,t) = 1$ for the WD-AK-111 coverings at sequence (W,t), that is, takes place additive action of factors. At sequence (t, W) factor $K_a(t, W) = 0.82$.

For the PVAC coverings, at sequence (W,t), value $K_a(W,t)$ is equal to 0.85, and at sequence (t, W) value $K_a(t, W)$ is equal to 0.72.

Thus, with identical working climatic factors and intensity, the value of the additivity coefficient for the various coverings can be different.

A decrease or increase in the humidifying time and thermal aging compared to the described in the given clause will obviously change the numerical values of the additivity coefficient, that is, an amplification or easing of the sequentially working factors can be observed.

Thus, during aging, the numerical values of the coverings' properties at the moment of time "t" are defined also by their kinetics of their previous changes, which depend on the intensity of the climatic influences (intensity of the UV irradiation, temperature, humidity of air, etc.), that is, the prehistories of aging. To account for the influence of the hereditary factor in the change of the coverings' properties, we will consider that at present time t the parameter of quality $U(t)$ represents the sum of two items:

$$U(t) = V(t) + \int_0^t V(\tau)K(\tau,t)d\tau \tag{4.34}$$

The first item represents the instant component, the second item the inherited component. It is defined as follows. The memoirs on a condition of coatings' properties at the moment of time τ and $\tau + d\tau$, belonging to the past, should be proportional to the size of property of a covering at the moment of time $\tau - V(\tau)$ and durations of an interval $d\tau$. For the account of a prehistory of aging, the function of forgetting $K(t,\tau)$ is used. If $K(t,\tau)$ aspires to the final limit at $t \to \infty$ and thus comes close to this limit quickly it is possible to show that the result of the previous influence is reduced to the addition of a decreasing component $\int_0^t V(\tau)K(t,\tau)d\tau$. Memory of it is kept forever, and thus in a covering there are irreversible changes.

In Eq. (4.34), $V(t)$ is the function characterizing the process of changing the properties of the coating without considering the hereditary factor. Function $K(t,\tau)$ characterizes the hereditary properties of a material and is a nucleus of heredity [7,11]. We represent the function $K(t,\tau)$ as a product of two functions

$$h(t,\tau) = h(\tau)\phi(t-\tau) \tag{4.35}$$

where $h(\tau)$ is the function describing the aging process of the coating.

Function $h(\tau)$ is usually approximated by the equation

$$h(\tau) = C_{\text{пр}} + C_o/\exp(-\beta\tau) \tag{4.36}$$

Constant $C_{\text{пр}}$ characterizes the limit of the coatings' properties.

$\phi(t-\tau)$ is the function describing the effect of duration of exposure, and characterizes the hereditary properties of the coating.

Function $\phi(t)$ changes within the limits of $0 < \phi(t) < 1$ $0 < t < \infty$.

After some mathematical transformations the model of aging of the coatings can generally be approximated by the expression

$$U(t) = A\exp(-\alpha t) - \alpha\int_0^t A\exp(-\alpha\tau)\{C_{\text{пр}} + C_0/\exp(-\beta t)\}(1-\exp(-(t-\tau)))d\tau \tag{4.37}$$

Numerical values of A, $\alpha, \beta, C_{\text{пр}}$ and C_0 in models of aging of some protective-decorative coverings received in view of the hereditary factor are obtained. Various numerical values of the functions with an influence of the hereditary factor on the change of the properties of coverings is established. The presence of a polymeric component in the structure of a covering promotes some decrease (reduction) in "memory" that is obviously caused by faster course relaxational processes in the structures of coverings. Numerical values of the function reflecting the influence of the hereditary factor depend on age, but the aspiration of a nucleus to a final limit with increase in age "t" is observed.

We have analyzed the change of the function characterizing the influence of hereditary factors on the properties of the coatings, depending on the type of coating and aging. We used the PVAC paint, and the polymer-lime and lime paint.

After curing, samples of coatings subjected to hydration and heat aging at 50°C were obtained. As an indicator of the quality we used the adhesive strength of coatings. The research results and calculations are given in Table 4.7.

For these types of coatings Eq. (4.37) has the form

1. **PVAC coatings**

   Thermal aging

$$R_{\text{сц}} = 1.97\exp(-0.0002t) - 0.0002\int_0^t 1.97\exp(-0.0002\tau)\{0.75 + 1.22/\exp(-0.0009t)\}\times(1-\exp(-(t-\tau)))d\tau$$

   Moistening

$$R_{\text{сц}} = 1.99\exp(-0.0003t) - 0.0003\int_0^t 1.99\exp(-0.0003\tau)\{0.75 + 1.24/\exp(-0.0015t)\}\times(1-\exp(-(t-\tau)))d\tau$$

2. **Lime paint**

   Thermal aging

$$R_{\text{сц}} = 1.1\exp(-0.003t) - 0.003\int_0^t 1.1\exp(-0.003\tau)\{0.1 + 0.99/\exp(-0.008t)\}\times(1-\exp(-(t-\tau)))d\tau$$

   Moistening

$$R_{\text{сц}} = 0.98\exp(-0.00085t) - 0.00085\int_0^t 0.98\exp(-0.00085\tau)\{0.25 + 0.75/\exp(-0.01t)\}\times(1-\exp(-(t-\tau)))d\tau$$

3. **Polymer-calcareous paint**

   Thermal aging

$$R_{\text{сц}} = 1.57\exp(-0.0005t) - 0.0005\int_0^t 1.57\exp(-0.0005\tau)\{0.75 + 0.84/\exp(-0.0004t)\}\times(1-\exp(-(t-\tau)))d\tau$$

   Moistening

$$R_{\text{сц}} = 1.66\exp(-0.0005t) - 0.0005\int_0^t 1.66\exp(-0.0005\tau)\{0.91 + 0.75/\exp(-0.003t)\}\times(1-\exp(-(t-\tau)))d\tau$$

Analysis of experimental data shows that the influence of the hereditary factor on the change of the durability of the adhesion of coverings is ambiguous. The maximal value of function $\int_0^t V(\tau)K(t,\tau)d\tau$ is typical of the limy covering in the process of thermal aging. After thermal aging for 200 hours the numerical values are 0.313 MPa for the lime coating. In the process of moistening, the effect of hereditary factors on the change in the strength of the adhesion of coatings were

Table 4.7 Value of the function that characterizes the influence of hereditary factors on the adhesion strength of some coatings, depending on the type of aging, MPa

Coating's type	Aging time (h)							
	25	**50**	**75**	**100**	**125**	**150**	**175**	**200**
PVAC paint	$\frac{0.018}{0.028}$	$\frac{0.037}{0.056}$	$\frac{0.056}{0.084}$	$\frac{0.074}{0.111}$	$\frac{0.092}{0.137}$	$\frac{0.109}{0.162}$	$\frac{0.127}{0.186}$	$\frac{0.144}{0.21}$
Polymer-calcareous paint	$\frac{0.029}{0.032}$	$\frac{0.057}{0.064}$	$\frac{0.084}{0.094}$	$\frac{0.11}{0.123}$	$\frac{0.133}{0.151}$	$\frac{0.156}{0.177}$	$\frac{0.178}{0.202}$	$\frac{0.198}{0.226}$
Limy paint	$\frac{0.077}{0.022}$	$\frac{0.139}{0.044}$	$\frac{0.187}{0.064}$	$\frac{0.225}{0.083}$	$\frac{0.225}{0.101}$	$\frac{0.279}{0.117}$	$\frac{0.298}{0.133}$	$\frac{0.313}{0.147}$

Note: Above the line the values of the function that characterizes the influence of hereditary factors on the properties of the coatings $\int_0^t V(\tau)K(t,\tau)d\tau$ during thermal aging are shown; below the line in the process of moistening.

affected to a lesser extent. After 200 hours of humidifying of a limy covering the value of function $\int_0^t V(\tau)K(t,\tau)d\tau$ is 0.147 MPa.

For the PVAC and limy coverings, the value of the function describing the influence of the hereditary factor on the change of the durability of the adhesion during humidifying in process of thermal aging is obviously caused by the stronger influence of a prehistory of aging in comparison with the destructive influence of temperature.

Taking into account the formation conditions of the coatings on the porous substrate, the coating has an initial level of accumulation of damages w_0. The value of the initial level of accumulation of damages among other factors has a significant effect on the change in the properties of the coatings during operation. To evaluate the effect of the initial level of damage accumulation on the hereditary component in the aging model of coatings (4.37), the following experiment was performed. Prepared cement substrates characterized by different porosity. The samples were painted with various paint. After curing which the samples were subjected to moistening-drying. As a criterion for the resistance of coatings, the adhesion strength R_c was determined, which was determined by the method of detachment of the washers. The results of the studies are given in Table 4.8.

Analysis of the experimental data (Table 4.8) shows, that with an increase in the initial level of accumulation of injuries w_0 is a higher value of the function, showing the influence of the hereditary factor is observed. Thus, after 150 cycles of moistening-drying of coating XB-161 the level of damage accumulation was $w = 0.612$, the value of the function $K\ (t,\ \tau) = 0.072$ at the initial accumulation level of damage was $w_0 = 0.4$ and while at the initial accumulation level damages $w_0 = 0.6$ these values are $w = 0.814$ and $K\ (t,\ \tau) = 0.09$, respectively. For coatings with a large initial level of accumulation of damages, a high rate of change in the function $K\ (t,\ \tau)$ is characteristic. Similar patterns are also characteristic for coatings based on WD-VA-17 paint.

Consider the process of aging of coatings from the position of accumulation of damages taking into account the hereditary factor. We will estimate the degree of damage to the material by the level of accumulation of damages W, which can be determined by formula

$$W = \frac{R_0 - R_\phi}{R_0} \tag{4.38}$$

where R_ϕ is the actual bond strength andR_0 is the adhesion strength of a defect-free coating.

Analysis of the experimental data (Fig. 4.12 and Table 4.9) shows that under the action of a constant voltage there is a proportional relationship between the logarithm of the adhesion failure time of the coatings and the index of the surface porosity of the substrate P. At increase of the surface porosity of the substrate the time of preservation of the adhesion contact “coating-substrate” decreases.

Table 4.8 Influence of the preliminary level of damage accumulation on the hereditary properties of coatings during the aging process

Test cycles	**0**	**25**	**50**	**75**	**100**	**125**	**150**
Porosity of the substrate 1.9%							
Level of accumulation of damages w	0.33 0,4	0.352 0.429	0.377 0.46	0.403 0.494	0.422 0.531	0.432 0.57	0.462 0.612
Hereditary factor	0 0	0,005 0.008	0.012 0.018	0.02 0.029	0.028 0.042	0.038 0.056	0.049 0.072
Porosity of the substrate 2.7%							
Level of accumulation of damages w	0.4 0.5	0.433 0.529	0.47 0.559	0.509 0.592	0.552 0.626	0.599 0.663	0.649 0.701
Hereditary factor	0 0	0.01 0.009	0.021 0.02	0.034 0.032	0.049 0.045	0.066 0.06	0.084 0.075
Porosity of the substrate 3.4%							
Level of accumulation of damages w	0.46 0.55	0.509 0.582	0.563 0.617	0.622 0.654	0.688 0.693	0.759 0.734	0.837 0.777
Hereditary factor	0 0	0.015 0.011	0.034 0.024	0.055 0.039	0.079 0.054	0.106 0.07	0.137 0.088
Porosity of the substrate 5.9%							
Level of accumulation of damages w	0.48 0.6	0.531 0.63	0.589 0.663	0.654 0.698	0.725 0.735	0.804 0.773	0.89 0.814
Hereditary factor	0 0	0.016 0.011	0.037 0.024	0.061 0.039	0.09 0.055	0.123 0.072	0.16 0.09

Note: Above the line are the values of the indicators for the coverage of WD-WA-17, below the line - for coating XV-161.

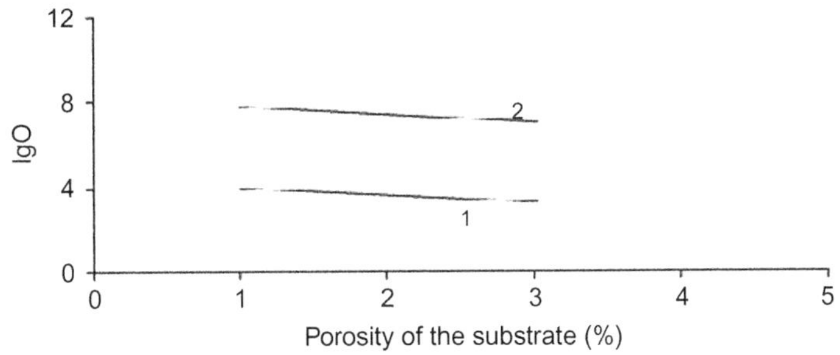

Figure 4.12 Effect of surface porosity of the substrate on the adhesion strength of coatings.

Table 4.9 Change of the adhesion strength of coatings depending on the porosity of the substrate

Kind of a coating		Adhesive strength (MPa)		
	Porosity of the substrate (%)	Before tests	After 50 test cycles	After 100 test cycles
Water dispersive WD-WA-17	1.9	2.2	1.9	1.75
	2.7	1.9	1.72	1.4
	3.4	1.79	1.27	1.03
	5.9	1.7	1.1	0.8
Perchlorovinyl XV-161	1.9	2.38	1.99	1.8
	2.7	1.9	1.78	1.4
	3.4	1.7	1.59	1.23
	5.9	1.57	1.26	0.8

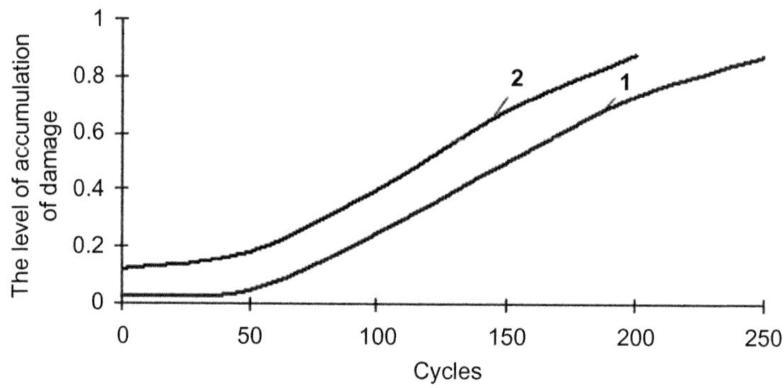

Figure 4.13 Changes in the level of accumulation of damage to the KO-168 coating during aging:
1—Coating on a substrate with a porosity of 12.3%
2—Coating on a substrate with a porosity of 23%

It is seen (Fig. 4.13) that at the beginning of the aging process (the incubation period), minimal change in the level of damage accumulation occurs.

The duration of the incubation period depends on the initial levelW_0. The greater the initial level of accumulation of damage W_0, the faster the rapid accumulation of damage begins, with the porosity of the cement substrate, $P = 12.3\%$ $W_0 = 0.04$, an incubation period of 25 cycles, with the porosity of the cement substrate, $P = 23\%$ $W_0 = 0.12$ and an incubation period of 15 cycles. Accordingly, the damage growth rate of the coating on a substrate with a porosity of 12.3% is lower than that of a coating on a substrate with a porosity of 23%. After 100 test cycles, the level of damage accumulation of the KO-168 coating on a substrate with a porosity of 12.3% is $W = 0.25$, and on a substrate with a porosity of 23% it is 0.37. At $W = 0.85$ the coatings are destroyed.

Using the approach of the hereditary theory of aging, we consider the function characterizing the change in the level of damage accumulation in the form

$$W(t) = W_0 \exp(\alpha t) + \int_0^t W_0 \exp(\alpha \tau) \frac{\partial}{\partial \tau} \left(\left(1 - \frac{W_1}{e^{\beta t}} \right) \times \left(1 - e^{-\gamma(t-\tau)} \right) \right) d\tau \quad (4.39)$$

The second term (we will denote it $K(t, \tau)$) is the effect of previous damages acquired during the operation of the coating up to this point in time.

Analysis of the calculated data shows that with an increase in the initial level of accumulation of lesions W_0, a higher value of the function, characterizing the influence of the hereditary factor is observed. Thus, after 150 cycles of humidification-drying of coating XV-161 the level of accumulation of damages was $W = 0.612$ and the value of the function $K(t, \tau) = 0.072$ at the initial level of damage accumulation was $W_0 = 0.4$, while at the initial level of accumulation of damage $W_0 = 0.6$ these values are $W = 0.814$ and $K(t, \tau) = 0.09$, respectively. For coatings with a large initial level of damage accumulation, the large rate of change of the function is characteristic $K(t, \tau)$. Similar patterns are also characteristic for coatings based on WD-WA-17 paint.

Thus, a more significant decrease in the adhesion strength of coatings on a substrate with porosity $P = 5.9\%$, (41%–53% depending on the type of coating) can be explained by the significant influence of the initial level of damage accumulation W_0 and of the hereditary factor of the kinetics after changing the properties of coatings.

The intensity of destruction under the influence of climatic factors is not the same at different stages of operation. This is mainly due to changes in the structure of the coating.

It has been established [12,13] that during a certain incubation period (the duration of which, depending on the type of coating, may be different), there is an insignificant increase in defectiveness. Thus, in the active stage of damage accumulation, significant changes occur in the structure and properties of the coating (color change, gloss, cracking, etc.).

In many cases, the kinetics of damage accumulation is described by the first-order differential equation

$$\frac{dW}{dt} = K(1 - W) \quad (4.40)$$

where k is the rate constant of damage accumulation.

A common solution is the function

$$W = 1 - Ce^{-kt} \quad (4.41)$$

If at the time of the start of operation due to structural defects in the coating there are already damages, that is, $W(0) = W_0$, then the desired dependence is a solution of the Cauchy problem in the form

$$W = 1 - (1 - W_0)e^{-kt} \tag{4.42}$$

In the incubation period and the active stage of damage accumulation, the rate of damage accumulation is different. For example, for PVAC coatings it is 0.00163 and 0.016 cycles^{-1}, respectively. For KO-168 coatings, the constant k in the incubation period is 0.00163 cycles^{-1}, and in the active stage 0.011 cycles^{-1}.

Similar conclusions can be drawn by the change in the aging process of the surface area of the coating. As is known, under the influence of climatic factors, the surface layers of a coating are destroyed, leading to an increase in the roughness, and consequently, the surface area of the coating. This increase in surface area is associated with the resulting macro- and microcracks.

Moisturizing is one of the aging factors. Analysis of the experimental data indicates that the beginning of the active stage of accumulation of lesions coincides in time with the first significant changes in the surface area of the coating (Figs. 4.14 and 4.15).

Experience shows that it can not be asserted that the intensity of the change in the properties of the coating is the same at different stages of the active stage of damage accumulation. Depending on the type of coverage, etc., the first half of

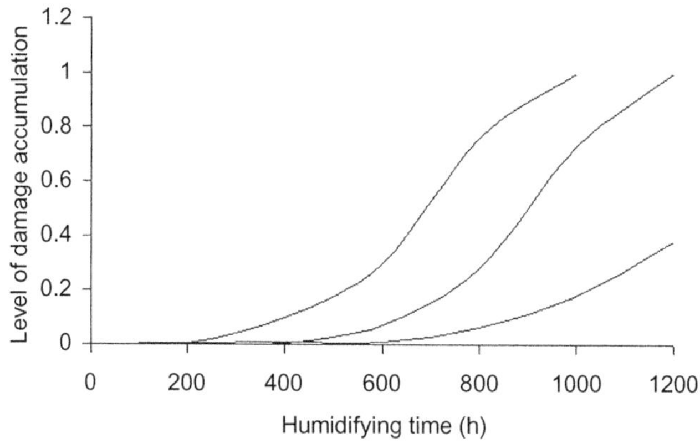

Figure 4.14 Changes in the level of accumulation of damage to coatings during the aging process.
1—PVAC coating
2—Polymer-calcareous coating
3—Lime coating

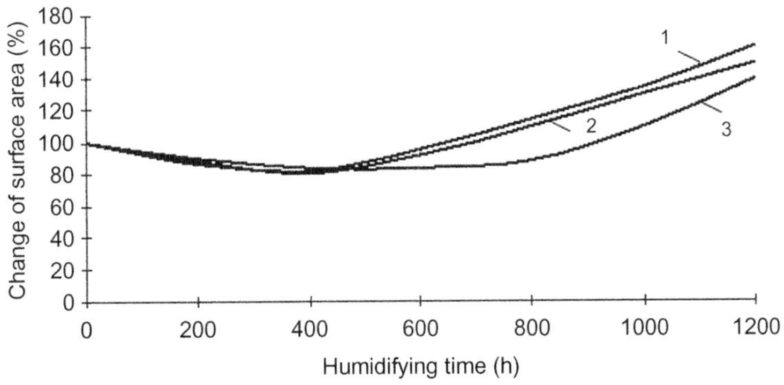

Figure 4.15 Changes in the surface area of coatings during humidification.
1—PVAC coating
2—Polymer-calcareous coating
3—Lime coating

the active stage is characterized by a particularly sharp increase in all negative parameters but later processes proceed more smoothly.

Thus, the kinetics of the destruction of coatings from a mathematical point of view should be described by differential equations of order higher than the first. This means that when analyzing a process, it is necessary to take into account not only the rate of change of the monitored parameter, but also at least the acceleration. Therefore, it is proposed to use, as a model of the fracture process

$$Y(t) = C_1 e^{\alpha_1 t} + C_2 e^{\alpha_2 t} + C_3 e^{\alpha_3 t} \tag{4.43}$$

which is a solution of a third-order differential equation.

The coefficients α_1, α_2, α_3 characterize the rate of change of the controlled parameter at different stages of coating aging. Analyzing the empirical curves for the kinetics of damage accumulation, we can assume that $\alpha_1 << \alpha_2$ (for the surface area of the coating, it is even characteristic $\alpha_1 < 0$), $\alpha_2 > \alpha_3$.

As an example, consider the change in the surface area of a PVAC coating during the humidification process.

From the experimental data it follows that $y(0) = 100$, $y(800) = 95$, $y(1200) = 140$, $y'(800) = 0$

Taking into account Eq. (4.43), we have to identify the parameters of the model:

$$\begin{aligned} &C_1 + C_2 + C_3 = 100 \\ &C_1 e^{800\alpha_1} + C_2 e^{800\alpha_2} + C_3 e^{800\alpha_3} = 95 \\ &C_1 e^{1200\alpha_1} + C_2 e^{1200\alpha_2} + C_3 e^{1200\alpha_3} = 140 \\ &\alpha_1 C_1 e^{800\alpha_1} + \alpha_2 C_2 e^{800\alpha_2} + \alpha_3 C_3 e^{800\alpha_3} = 0 \end{aligned} \tag{4.44}$$

From the first three equations we obtain

$$C_1 = \frac{100e^{800\alpha_2+1200\alpha_3} - 100e^{800\alpha_3+1200\alpha_2} - 95e^{1200\alpha_3} + 140e^{800\alpha_3} + 95e^{1200\alpha_2} - 140e^{800\alpha_2}}{e^{800\alpha_2+1200\alpha_3} - e^{800\alpha_3+1200\alpha_2} - e^{800\alpha_1+1200\alpha_3} + e^{800\alpha_3+1200\alpha_1} + e^{800\alpha_1+1200\alpha_2} - e^{800\alpha_2+1200\alpha_1}}$$

$$C_2 = \frac{95e^{1200\alpha_3} - 140e^{800\alpha_3} - 100e^{800\alpha_1+1200\alpha_3} + 100e^{800\alpha_3+1200\alpha_1} - + 140e^{800\alpha_1} - 95e^{1200\alpha_1}}{e^{800\alpha_2+1200\alpha_3} - e^{800\alpha_3+1200\alpha_2} - e^{800\alpha_1+1200\alpha_3} + e^{800\alpha_3+1200\alpha_1} + e^{800\alpha_1+1200\alpha_2} - e^{800\alpha_2+1200\alpha_1}}$$

$$C_2 = \frac{95e^{1200\alpha_3} - 140e^{800\alpha_3} - 100e^{800\alpha_1+1200\alpha_3} + 100e^{800\alpha_3+1200\alpha_1} - + 140e^{800\alpha_1} - 95e^{1200\alpha_1}}{e^{800\alpha_2+1200\alpha_3} - e^{800\alpha_3+1200\alpha_2} - e^{800\alpha_1+1200\alpha_3} + e^{800\alpha_3+1200\alpha_1} + e^{800\alpha_1+1200\alpha_2} - e^{800\alpha_2+1200\alpha_1}}$$

We also determine α_1 and α_3 from the experimental characteristic $Y\,(t)$ (Fig. 4.16). Since $\alpha_1 < < \alpha_2$, the component $y(t+T) = Ae^{-\lambda_2(t+T)}$ determines the process until the end of the incubation period, that is, at $0 \le t \le 800$. The value can be determined at the end of the experimentally obtained process $Y\,(t)$.

$Y(0+T) = Ae^{\alpha_1(0+T)}$, at $t < 800$,

$$y(t+T) = Ae^{-\lambda_2(t+T)}, t > > 800,$$

$$\alpha_1 = \frac{\ln(Y(T)/Y(0))}{T},$$

$$\alpha_3 = \frac{\ln(Y(t+T)/Y(t))}{T}.$$

In this case, $\alpha_1 = -0.000064$, $\alpha_3 = 0.0012$.

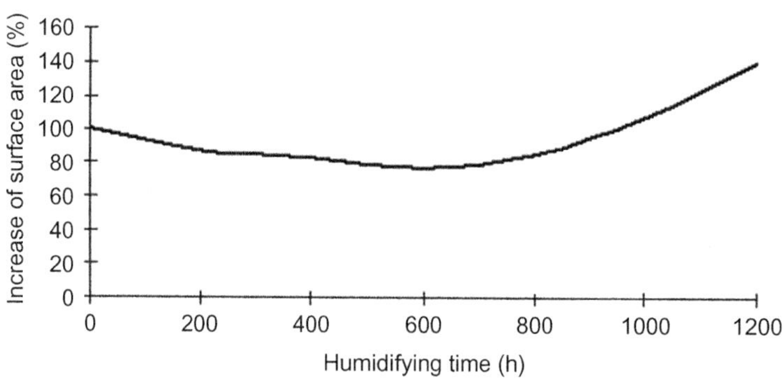

Figure 4.16 Changes in the surface area of PVAC coatings during the aging process.

From the fourth equation of system (4.44) we obtain an implicit expression for α_2:

$$-0.00006C_1 + \alpha_2 C_2 e^{800\alpha_2} + 0.003C_3 = 0 \tag{4.45}$$

An approximate solution of this equation gives $\alpha_2 = 0.011$.

As a result, the dependence Eq. (4.43) is represented in the form

$$Y(t) = 100,364e^{-0.000064t} + 0.00009e^{0.011t} - 0.364e^{0.0012t}.$$

From the above, the algorithm for identifying kinetic processes of a given type follows:

1. According to the initial changes in the incubation period α_1, it is determined, at the end of the empirical curve, α_3.
2. Constants C_1, C_2, C_3 are represented as functions of α_2.
3. On the characteristic point of the empirical curve, Eq. (4.43) is constructed, and an approximate value of the constant α_2 is found.
4. are identify

$$Y(t) = C_1 e^{\alpha_1 t} + C_2 e^{\alpha_2 t} + C_3 e^{\alpha_3 t} \tag{4.46}$$

The proposed algorithm can be used to solve other problems in building materials science with the possibility of describing the processes under consideration as solutions of differential equations of the third order.

Considering on the basis of this function the hereditary component

$$I(t) = \int_0^t K(t,\tau)\frac{dY(\tau)}{d\tau} \tag{4.47}$$

we can conclude that in the stages of aging of the coating described above, it is constructed additively for each period separately. In the incubation period, its influence is positive; in the active stage, it aggravates the aging process.

Carrying out calculations for other coatings, we have (Table 4.10).

Applying the principles of the hereditary theory of aging to these dependences, it is possible to calculate the influence of the hereditary factor on the change in the surface area of the coatings after moistening (Tables 4.11–4.13).

Table 4.10 Model of the fracture process of coating

Lime coating	$Y(t) = 129.292e^{-0.00017t} + 14.06e^{0.002t} - 43.325e^{0.0007t}$
Polymer-calcareous coating	$Y(t) = 110.059e^{-0.00021t} + 2.089e^{0.003t} - 12.898e^{0.0091t}$
PVAC coating	$Y(t) = 100.364e^{-0.000064t} + 0.00009e^{0.011t} - 0.364e^{0.0012t}$

Table 4.11 Share of change (%) due to the hereditary factor in the total change in the surface area of the lime coating during the moistening process

The lime coating	Time (h)	I	II	III
	100	**− 0.099**		
	200	**− 0.386**		
	300	**− 0.846**		
	400	**− 1.467**	**31.562**	
	600		**44.961**	
	800		**50.708**	**29.638**
	1000			**39.205**
	1200			**47.885**
	The share of change due to the hereditary factor in the total change for phase	−42%	+28%	+7.8%

Table 4.12 Share of change (%) due to the hereditary factor in the total change in the surface area of the polymer-calcareous coating during the moistening process

The polymer-calcareous coating	Time (h)	I	II	III
	100	**− 0.086**		
	200	**− 0.337**		
	300	**− 0.738**		
	400	**− 1.278**	**6.327**	
	600		**7.259**	
	800		**6.594**	**10.83**
	1000			**13.417**
	1200			**15.347**
	The share of change due to the hereditary factor in the total change for phase за фазу	−25,4%	66,9%	8.9%

Table 4.13 Share of change (%) due to the hereditary factor in the total change in the surface area of the PVAC coating during the moistening process

PVAC coating	Time (h)	I	II	III
	100	**− 0.026**		
	200	**− 0.101**		
	300	**− 0.221**		
	400	**− 0.381**		
	600	**− 0.809**		
	800	**− 1.359**	**0.039**	
	1000		**8.171×10^{-3}**	**10.554**
	1200			**13.141**
	The share of change due to the hereditary factor in the total change for phase	−30.6%	48.1%	2.9%

4.2 Prediction of the aging time of coatings of exterior walls of buildings

Examination survey results of painted facades of buildings show that the actual service life does not always correspond to the planned service life, which leads to additional costs for the repair of building facades. The current empirical assessment of service life can not produce reliable data on the durability of coatings, due to the lack of theoretical development to assess the service life.

There are several approaches to forecasting the durability of coatings. In Ref. [4] it was proposed to evaluate the durability of coatings with regard to the internal stresses. However, destruction of coverings is caused not by one kind of influence such as voltage, but also humidity, solar illumination and other factors. In Ref. [3] it is suggested to estimate the durability of coatings τ_H using the results of accelerated tests with use of linear dependence τ_H on the duration of the accelerated tests τ_y. However, at application of various paints the linear model has a various kind. It reduces reliability of forecasting.

Analysis of the scientific and technical literature shows that questions of forecasting remain. Thus, creating an evidence-based method for predicting service life is an important scientific, technical and economic problem.

The aim is to develop a methodology for assessing the life of coatings exterior walls.

Justification of the complexity of the mathematical model is a difficult part of aging coatings. In the operation of protective and decorative coatings exterior walls, the character of the aging mechanism at different stages of aging determine different components of the mechanism of aging. It may be chemical processes in coatings, occurring as a continuation of the curing process, external influences only change their speed, is the chemical processes in coatings, including on the surface of pigments and fillers, which are the result of external influences the reactants (oxygen acids, alkalis, water in the case of the hydrolysis, etc.) and activating factors (light, temperature) (e.g., hlorvinilovye polymers), physicochemical processes leading to structural changes (supramolecular structure and phase).

However, in most cases, the destruction is the result of just a few processes. Consider the aging process from the perspective of the coating thermo-fluctuation theory of strength polymers [7,14]. According to the molecular model of fracture atoms in the tip of a crack due to thermal fluctuations from time to time acquire enough kinetic energy to rupture or recovery. Under the influence of climatic factors in the coating of internal stresses, the probability of breaking the bonds in the structure of the coating increases, and their recovery is reduced. According to the molecular model, the value of the potential fracture energy needed to break the bond is

$$U = U_o - v_a \sigma^x \tag{4.48}$$

where U_o is the potential barrier being torn down,σ^x is the action local stress andv_a is the amount of fluctuation.

The magnitude of the potential energy needed to restore communication is

$$U* = U_o + v\sigma^x \tag{4.49}$$

where $U*$ is a potential barrier recombination connection with $\sigma = 0$.

At some value of stresses in the coating under the influence of climatic factors $U_o - v\sigma^x = U_o^* + v\sigma^x$, the crack does not grow, but there is a slight change in the decorative and protective properties (incubation period). At voltages $\sigma > \sigma^x$ observed crack growth, the speed of the process v is determined by the relation

$$v = v_1 - v_2 \tag{4.50}$$

where v_1 is the crack growth rate and v_2 is the velocity of the fracture closure.

As $v_1 > v_2$ the rate of aging is determined by the rate of crack growth. The frequency of fluctuations in time of disconnection at the crack tip is

$$u = u_o \exp[-(U_o - v\sigma^x)/kT] \tag{4.51}$$

Since the crack growth rate is $v_1 = \lambda u_1$, in view of formula (4.51)

$$v - \lambda u_o \exp[-(U_o - v\sigma^x)/kT] \tag{4.52}$$

The transition from the active to the incubation period is observed at step $\sigma > \sigma^x$.

The lifetime may be defined by the formula

$$\tau = \int_{w_o}^{w_k} \frac{dw}{v} \tag{4.53}$$

where w_o, w_k is the initial and final level of damage accumulation.

The solution of Eq. (4.51) leads to the form

$$\tau = A\tau_o \exp[(U - v_\sigma)/RT] \tag{4.54}$$

Expression is $(U - v_\sigma)$ the activation energy of aging. It is an effective value and a combination of the activation energies of individual processes underlying aging coatings. The factor is variable depending on the humidity, the intensity of UV radiation.

Consider the effect of humidity and UV radiation by a factor. As is known

$$\beta = \gamma W \tag{4.55}$$

where γ is the coefficient of proportionality, taking into account the effect of humidity on the rate of hydrolysis of the film-forming agent; and $W$is the air humidity.

The rate of hydrolysis is proportional to the amount of adsorbed moisture, and can be described by the reaction equation of the first order

$$\beta = (1/t)\ln\{C_o/C\} \tag{4.56}$$

where C is the concentration of the substance.

The ratio of the aging time when changing humidity will describe the dependence:

$$\tau_1/\tau_2 = \{(1/\gamma W_1)\ln(C_o/C_1)\}/\{(1/\gamma W_2)\ln(C_o/C_2)\} = \text{const } k_1 \tag{4.57}$$

The duration to achieve the same degree of fracture based on Eq. (4.57) can be determined by

$$\tau_1 = \tau_2 k_1 \tag{4.58}$$

The minimum humidity value achieved in the area of operation is used as an example [6,7].

Table 4.14 shows the results of experimental and calculated data on the effect of humidity on the length of aging coatings on the silicone paint KO-168. The calculated data show good agreement with the experimental data. In the calculation of the confidence probability of 0.95, the critical value of pair correlation coefficients for the 95% confidence probability and the number of degrees of freedom $n = 5$ was $= 0.707$. The calculated value of the coefficient of pair correlation $= 0.902$ was that much more critical and shows a high consistency between the experimental and calculated data.

In Refs. [15–18], the results of the studies of kinetics of degradation of coatings under the action of light radiation are given. The duration of the test to the same degree of damage depends linearly on the back radiation intensities

$$\tau_1 J_1 = \tau_2 J_2 = \tau_3 J_3 = \text{const} = k_2 \tag{4.59}$$

Table 4.14 Duration of aging coatings KO-168 with the loss of gloss 20%

Type of substrate	**The porosity of the substrate (%)**	**Humidity (%)**			
		60		**100**	
Without UV irradiation					
Glass	0	1700/1698	941/933	1346/1350	727/724
Porous cement	20.3	914/912	515/517	701/706	396/401

Note. Above the line shows the experimental data duration of the test; below the line the calculated data.

Given Eq. (4.58), the duration of the aging to a predetermined degree of coating damage depends on the radiation intensity determined by the relation

$$\tau_1 = \tau_2 k_2 \tag{4.60}$$

The duration of the aging of coatings to the desired degree of destruction is determined by the ratio

$$\tau = \tau_o \exp[U/RT]k_1 k_2 \tag{4.61}$$

With variable operating conditions $x(t)$, use the Bailey equation

$$\int_o^T dt/\tau = 1 \tag{4.62}$$

where t is the durability of the coating operation under these conditions and τ is the coating durability under all operating conditions.

In the transition from the integral form for a finite increment

$$\sum \Delta t/\tau = 1 \tag{4.63}$$

The methodology for calculating the life of coatings used the additivity principle that allows you to define constant operating conditions equivalent to the total of their destructive effects to various conditions of exploitation.

As constant conditions of exploitation: temperature –273K; minimal achievable humidity; of intensity UV radiation in this climatic region. Equivalent time proposed to determine the formula

$$\tau_{\text{ЭКВ}} = (W_{\min}/W_{\text{ТЕК}})(J_{\min}/J_{\text{ТЕК}})\tau_o \exp[-U/R(1/T_{мек} - 1/T_o)] \tag{4.64}$$

where $T_{\text{ТЕК}}$ is the current operating temperature.

The algorithm for determining the duration of the aging of coatings to the desired degree of destruction is as follows.

1. Determine for a given climatic region, in accordance with the mathematical model (4.61) time during the year, which is equivalent to the total destructive effect with respect to the constant parameters (temperature, humidity, radiation intensity). You must first determine the value of the effective activation energy U.

Determine the intensity of climatic tests with the regime and the number of test cycles

$$t_2 = N\tau_{экв} \tag{4.65}$$

where N is the number of test cycles and $\tau_{экв}$ is the intensity of one cycle of climatic testing per day.

Determine the lifetime of the formula

$$\tau = \frac{t_2}{t_1} \tag{4.66}$$

The following is an example of calculating lifetime PVAC (ПВАЦ) and silicone (KO-166) coatings. Calculations were made using the average monthly value of the intensity of UV radiation with wavelengths less than 400 nm, the relative humidity for a moderately cold climate. Preliminary studies and calculations showed that the activation energy of the ПВАЦ coating is $U = 92{,}230$ kDzh/mol, covering KO-166-94.54 kJ/mol. Accelerated testing was performed with 4 hours of freezing at −40°C; 2 hours of thawing in air at a temperature of 40°C and a relative humidity of 60%; 2 hours of moisture at a temperature of +20°C; 16 hours - UV radiation at a temperature of +20°C and a relative humidity 70%. The results of the calculations are given in Table 4.15.

Polyvinyl-acetate-cement coating

When the air temperature reaches −40°C during tests, the equivalent time at 0°C is

$$\tau_{экв} = \frac{4}{24}\exp\left(\frac{92,230}{R} : \left(\frac{1}{273} - \frac{1}{233}\right)\right) = 0.00013 \text{ day}$$

When the air temperature of 20°C and a relative humidity of 100% is equivalent to the time

Table 4.15 Equivalent $\tau_{экв}$coating operation with respect to 0°C in different climatic zones

Month	Moscow	Yakutsk	Vladivostok
1	2	3	4
January	0,61/0,58	0,001368/0,001125	0,34/0,323
February	0,695/0,67	0,0065/0,0056	0,56/0,54
March	5,64/5,54	0,318/0,288	6,74/6,65
April	28,525/19,34	15,86/15,85	28,52/28,92
May	90,32/94,2	39,41/40,2	98,87/75
June	191,32/202,29	169,38/178,32	131,30/137,35
Jul	300,99/320.63	325,05/346,80	278,54/295,85
August	181,00/191,65	149,95/157,95	294,63/315,79
September	51,0/56,5	29,97/30,26	111,92/116,74
October	3,05/12,41	1,93/1,88	29/30,42
November	2,44/2,4	0,0327/0,029	2,78/2,78
December	0,664/0,65	0,002/0,0017	0,492/0,47
In total	860,35/906,86	716,25/771,6	871,75/1010,84

Note. Above the line shows the values for PVAC coating; below the line coating KO-166.

$$\tau_{экв} = \frac{2}{24} \times \frac{100}{90} \exp\left(\frac{92,230}{R} : \left(\frac{1}{273} - \frac{1}{293}\right)\right) = 1.58 \text{ day}$$

When the air temperature + 20°C. 70% relative humidity and exposure to UV radiation equivalent time is

$$\tau_{экв} = \frac{16}{24} \times \frac{70}{90} \times \frac{912}{333} \exp\left(\frac{92,230}{R} : \left(\frac{1}{273} - \frac{1}{293}\right)\right) = 22.58 \text{ days}$$

Therefore, a test cycle is equivalent to 34.16 days at 273K. The number of test cycles at 120

34.16 × 120 = 4099.84 days.

The service life of coatings is equivalent to

- Moscow

$$\frac{4099.84}{860.35} = 4.78 \text{ years}$$

- Yakutsk

$$\frac{4099.84}{716.25} = 5.72 \text{ years}$$

- Vladivostok

$$\frac{4099.84}{871.75} = 4.7 \text{ years}$$

Organicsilicon coating KO-168

When the air temperature is −40°C during tests, equivalent time at 0°C is

$$\tau_{экв} = \frac{4}{24} exp\left(\frac{94,500}{R} : \left(\frac{1}{273} - \frac{1}{233}\right)\right) = 0.00013 \text{ day}$$

When the air temperature −40°C and a relative humidity of 60% is equivalent to the time

$$\tau_{экв} = \frac{2}{24} \times \frac{60}{90} exp\left(\frac{94,500}{R} : \left(\frac{1}{273} - \frac{1}{313}\right)\right) = 11.51 \text{ days}$$

When the air temperature of 20°C and a relative humidity of 100% is equivalent to the time

$$\tau_{экв} = \frac{2}{24} \times \frac{100}{90} exp\left(\frac{94,500}{R} : \left(\frac{1}{273} - \frac{1}{293}\right)\right) = 1.58 \text{ day}$$

One test cycle is equivalent to 37.32 days at a temperature of 273K. The number of test cycles is 200

$37.32 \times 200 = 7465.59$ days. The service life of coatings is equivalent to

- Moscow

$$\frac{7465.59}{906.86} = 8.24 \text{ years}$$

- Yakutsk

$$\frac{7465.59}{771.6} = 9.65 \text{ years}$$

- Vladivostok

$$\frac{7465.59}{1010.84} = 7.38 \text{ years}$$

Results of field tests confirmed that the discrepancy between the predicted and the real life does not exceed 15%.

References

[1] M.I. Karjakin, Physico-Chemical Basis of the Formation and Aging of the Coatings, Chemistry, Moscow, 1980, p. 216.

[2] M.I. Karjakina, Test Cover and Paints, Chemistry, Moscow, 1990 [4] L.P. Orentlikher, V.I. Loganina, S.I. Mishanin, Metodika life prediction coatings, Proceedings of the Universities. Building. 9 (1994), 22–23.

[3] L.A. Suhareva, Durability of Polymer Coatings, Chemistry, Moscow, 1984, p. 240.

[4] P.I. Zubov, L.A. Suhareva, Structure and properties of polymer pokrytiy, Chemistry, Moscow, 1982, p. 256.

[5] E.A. Barbashev, et al., Effect of radiation and heat aging resistance, fatigue and microstructure fiberglass UPU-7, Physicochem Mech. Polimerov N6 (1989) 68–71.

[6] V.I. Loganina, O.V.Karpova, By assessing the kinetics of aging coatings // Proceedings of the universities. Building. 2 (1998).

[7] JU. N Rabotnov, Mechanics of a Deformable Body, Science, Moscow, 1988.

[8] G.M. Bartenev, Strength and Fracture Mechanism Polimerov, Chemistry, Moscow, 1984, p. 280.

[9] I.E. Prokopovich, V.A. Zedgenidze, Applied Theory of Creep, Stroyizdat, Moscow, 1980, p. 280.

[10] N.M. Sedyakin, On a physical principle reliability, News of an Academy of Sciences USSR, Technical Cybernetics, N3, 1966.

[11] V.A. Smagie, About one model of the forced tests, Reliability and Quality Assurance, 1966, N4.

[12] L.P. Orentlikher, V.I. Loganina, On the question of the destruction of cement concrete coatings, Housing 8 (1999).

[13] V.I. Loganina, On the question of predicting the lifetime of protective and decorative coatings cement concrete, Proceedings of the Universities. Building 3 (1996).
[14] S.N. Zhurkov, B.N. Nazrullaen, Temporary dependence the strength of solids, J. Techn. Phys. 23 (10) (1983) 1677.
[15] E.A. Andryushchenko, Light Resistance of Coatings, Chemistry, Moscow, 1986, p. 187.
[16] V.I. Loganina, L.V. Makarova, S.N. Rislithyna, Assessment of an aging of protective decorative coatings, Contemp. Eng. Sci. 7 (36) (2014). Available from: https://doi.org/10.12988/ces.2014.411213. 61-1965HIKARI Ltd, www.m-hikari.com.
[17] V.I.Loganina, The Kinetics Model of Coverings' Properties with Consideration of the Heredity Factor Contemporary Engineering Sciences, 8, 2015, 2, 85-89HIKARI Ltd, www.m-hikari.com https://doi.org/10.12988/ces.2015.412257
[18] V.I. Loganina, Model for predicting protective and decorative exterior walls, Contemp. Eng. Sci. 8 (2) (2015). Available from: https://doi.org/10.12988/ces.2015.412251. 77-84HIKARI Ltd, www.m-hikari.com.

Statistical methods of quality management of coatings of cement concrete

5

5.1 Application of the six sigma methodology to the analysis of the quality of paint and varnish coatings

Construction and maintenance of the working condition of existing buildings and new constructions require a variety of paint and varnish structures. Growing competition in the business of finishing materials has increased consumer demand and requires manufacturers to produce materials for high-quality painted surfaces. However, currently manufacturing processes resulting in low-quality finishes are common, resulting in unplanned repairs and additional expenses due to the low-quality materials being used in manufacturing.

Any indicator characterizing the quality of products, as a rule, has one-sided and two-sided tolerances with specified values (UCL and LCL referring to upper and lower tolerance limits). going beyond the tolerance limits leads to marriage. the more standard deviation σ of the process fits in the tolerance field, the fewer defects will be produced in production [1–4]. Table 5.1 shows the probability of a random value over the tolerance limits for different values of root-mean-square deviations in the tolerance zone.

An excellent indicator is the hit $(3-4)\sigma$ in the interval between the mean value and the control limit. this means the presence of 6200 defects per million products. the term "six sigma" arises when trying to achieve such a process variability that $\pm 6\sigma$ (the standard deviation estimate) is between the upper and lower tolerance limits for the process.

Table 5.1 Probability of output of a random value over the limits of tolerance $(m \pm k\sigma)$

ε/P	An arbitrary distribution law	Normal distribution law	
		Two-tail position	One-tail position
2σ	$25,000 \times 10^{-6}$	$45,000 \times 10^{-6}$	$22,750 \times 10^{-6}$
3σ	$111,100 \times 10^{-6}$	2699×10^{-6}	1395×10^{-6}
4σ	$62,500 \times 10^{-6}$	$63,372 \times 10^{-6}$	$31,686 \times 10^{-6}$
5σ	$40,000 \times 10^{-6}$	0.5742×10^{-6}	0.28715×10^{-6}
6σ	$27,700 \times 10^{-6}$	0.00198×10^{-6}	0.001×10^{-6}

Increasing the Durability of Paint and Varnish Coatings in Building Products and Construction.
DOI: https://doi.org/10.1016/B978-0-12-817046-5.00005-X

Consider the application of the six sigma methodology applied to the production process. The level of defectiveness can be determined in accordance with the formula

$$1 - \frac{\text{Number of defective units}}{\text{Total number of products}} \tag{5.1}$$

Fig. 5.1 shows the "ideal" process (curve 1), when the mean and standard deviation of quality indicators provide an ideal quality, and the "critical" process (curve 2), in which even the slightest deterioration in parameters or an increase in the standard deviation will lead to an excess of the permissible defect rate.

Suppose that the process deviated from the average by 1.5σ and that the number of products is 100,000. Then the number of products whose quality indicators will be below the lower control limit will be 66.807. Thus it is necessary to strive for a variance $\pm 6\sigma$ fits within the interval from the lower control limit to the mean. In this case, even the displacement of the process will not lead to the appearance of a large number of defects. In this case, even if the process average shifts by 1.5σ (e.g., by $+1.5$ sigma closer to the upper boundary), the number of defects remains very small and amounts to only 3.4 defects per million possibilities.

When we evaluate the quality of the final product, the low value of the defectiveness level at first glance indicates the efficiency of the production process. However, if we consider the internal indicators of defectiveness, we can present in quantitative terms the amount of improvements during the process. Consider the analysis of the defectiveness of the process on the example of obtaining protective and decorative coatings for building products and structures. In accordance with TR 140-03, "technical recommendations on the technology of painting interiors and facades of residential and public buildings under construction," the process of painting involves the following operations:

- surface cleaning;
- priming of the cleaned surface and hardening of weak crumbling bases;

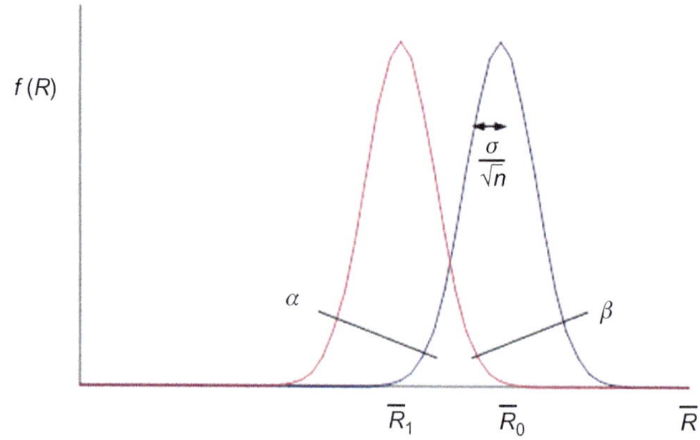

Figure 5.1 The laws of distribution of average values of the quality index of "standard" and "substandard" products.

- shpatlevanie:
  - padding;
  - painting of the first layer of paint;
  - painting a second coat of paint.

Suppose that the data at the output of the process indicate that the final output of high-quality colored products is 98%, that is, of the 1000 painted products 980 products are painted without marriage, and 20 products are rejected (defectiveness level value is 2%).

As already noted, the process of painting works consists of six operations, each of which works with a certain level of defectiveness. Suppose that in the process of operational control, defective products were identified, which were sent for revision. In Table 5.2, the values of the quality level at each stage of the process are given.

Analysis of the data (Table 5.2) shows that the number of units sent for revision is 84. The actual quality of the process is as follows:

$$1 - \frac{84 \text{ revision units}}{1000} = 0.916 = 91.6\%$$

Therefore the quality level value of 98% does not reflect the real picture of the quality of the process.

This discrepancy between the quality indicators of 98% and 91.6% is due to the fact that the figure of 98% "hides" the defects that are eliminated during the process. That is, 84 units were sent for revision, but 64 of them were corrected and returned to the staining process.

Based on the normal distribution law, the standard deviation of the quality index is equal to:

Table 5.2 Process quality

No.	Name operations	Number of units at the entrance	Number of units at the exit	Number of units for revision	Process quality (%)
1	Surface cleaning	1000	970	30	97.0
2	Priming of a cleaned surface and hardening of weak crumbling bases	970	950	20	97.93
3	Application of putty	950	935	15	98.42
4	Padding	935	930	5	99.46
5	Painting of the first coat of paint	930	926	4	99.57
6	Painting of the second coat of paint	926	916	10	98.92

Table 5.3 Number of standard deviations

No.	Name operations	Number of standard deviations σ
1	Surface cleaning	2.17σ
2	Priming of a cleaned surface and hardening of weak crumbling bases	2.3σ
3	Putty	2.42σ
4	Padding	2.78σ
5	Painting of the first coat of paint	2.86σ
6	Painting of the second coat of paint	2.54σ

- the first option (calculation based on the final result) -2.33σ and
- the second option (considering the level of defectiveness at each stage of the process) -1.73σ

Analysis of the data (Tables 5.2 and 5.3) shows that in terms of the number of standard deviations σ and the level of defectiveness, the greatest attention is required by the surface cleaning operation, the priming of the cleaned surface and hardening of weak bases, as well as the filling (putty) of the surface [5–8].

Suppose that products have dimensions of $1.5 \times 6\ \text{m}^2$. In accordance with TEP 81-02-15-2001 in the Penza region, the damage caused by the fact that 20 products (with an area of $180\ \text{m}^2$) were found to be defective:

$$У = З \times K_{\text{index}} \times S = 1830.48 \times 4.491 \times 1.8 = 14,797.24 \text{ ruble}$$

where $У$ is the damage (less profit) ruble; K_{index} is the indexation coefficient, which is 4491 (with VAT and materials); З is the direct costs, including payment for labor, materials, and operation of machines; and S is the increase in area compared to $100\ \text{m}^2$.

Suppose the defects were committed:

- when the surface is putty (15 units were sent for revision) and
- when cleaning the surface, priming the cleaned surface and hardening of weak crumbling bases (20 units were sent for revision).

In addition, direct costs are:

- with putty (the area of 15 products is $135\ \text{m}^2$) and
- when cleaning the surface, priming the cleaned surface and hardening the weak crumbling grounds 20 products with an area of $180\ \text{m}^2$.

$$ДЗ = З \times K_{\text{index}} S = 3600 \text{ ruble}$$

Thus additional direct costs are

$$ДЗ = 2264.9 + 3600 + 14,797.24 = 20,669.14 \text{ ruble}$$

Consideration of economic costs at each stage of the process allows the possibility of increasing production efficiency. To reduce financial losses the number of standard deviations of quality indicators in the tolerance field must be increased, as is recommended by the six sigma methodology. To successfully reduce the cost of correcting defects, it is necessary to carry out preventive and corrective actions, as well as to evaluate the failures leading to a decrease in profits, regardless of the cause.

To assess the potential performance of the process used the index of reproducibility C_{pk} and the process capability index are calculated by the formula [9–11]:

$$C_{pk} = \min\left\{\begin{array}{c}\frac{\mathrm{UT}-\bar{x}}{3\sigma}\\ \frac{x-\mathrm{LT}}{3\sigma}\end{array}\right\} \tag{5.2}$$

where σ is the standard deviation; $\bar{x}$ is the average value of the quality measure; and UT and LT are the upper and lower tolerance on quality, respectively. When $C_{pk} < 1.0$ the process is not reproducible; when $C_{pk} = 1.0$ the process is reproducible, but requires careful attention; and when $C_{pk} > 1.0$ the process is reproducible.

The Japanese expert on statistics Taguchi has suggested characterizing product stability using quality indicators as the so-called functions of losses that simultaneously consider losses both on the part of the consumer and on the part of the manufacturer [12–14].

The loss function has the following form:

$$L(y_i) = k(y_i - y_o)^2 \tag{5.3}$$

where L is the losses for a society (with size taking into account losses of the consumer or the manufacturer due to defective production); k is the constant of losses determined in view of charges of the manufacturer of products; y is the value of the measured functional characteristic; and y_o is the target value of the examined characteristic.

If you consider that the level of quality aggregate consists of n units, the additional costs borne by the consumer or the manufacturer can be defined by the formula

$$L = kd^2 \tag{5.4}$$

where d^2 is the size equal.

$$d^2 = \frac{1}{N}\sum_{i}^{i=N}(y_i - y_o)^2 = \sigma^2 + (\bar{y} + y_o)^2 \tag{5.5}$$

Considered as the mean square deviation from the target characteristics y defining the quality of some population units.

In accordance with Eq. (5.5) the loss is determined by two sources of variation: the middle $\bar{y}$ position relative to the target values y_o and the spread of values around its middle $\bar{y}$ characteristics.

Let us consider the financial expenses of a manufacturer of poor quality painting works, depending on reproducibility of process in accordance with Eq. (5.3) low loss if the process is fine-tuned (arithmetic mean value coincides with the nominal y_o).

At analysis of the process of staining paint PF-115 was assessed of the coating surface roughness [15]. To evaluate the uniformity of distribution of the roughness parameters were calculated along the strike of the arithmetic mean $\bar{y}$, standard deviation σ. consider the quality of the painted surface, which is characterized by a class N5. The tolerances for class roughness N5 in accordance with ISO 1302 N5 are: LT = 0.4 mkm and UT = 16 mkm. Fine-tuning the dyeing process for the class roughness N5 implies target y_o = 0.4 mkm equal.

According to the cost of repair painting works ranging from 573 to 1219.54 rubles (on 100 m^2), depending on the type of paint composition and production technology of painting works. Suppose that the manufacturer bears the cost of repairing the painted surface due to the lower quality of the appearance of the coating (increased roughness on 0.5 mkm). In accordance with Eq. (5.4) economic constant k will be

$$k = \frac{1219.54}{0.25} = 4878.16\ \text{ruble/mkm}^2$$

The results of calculations of statistical indicators of the quality of the painted surface and additional financial costs associated with poor quality of finishing are shown in Tables 5.4 and 5.5.

Analysis of the data in Table 5.5 shows that an increase of an index of reproducibility C_{pk} from 0.579 up to 1.267 does not correlate with parameters of additional

Table 5.4 Statistical characteristics of coloring process

Method of application of paint	Arithmetical mean $\bar{y}$ (mkm)	Standard deviation σ (mkm)	Index of reproducibility C_{pk}	Average standard deviation y from the target characteristics, d^2
Brush	4.34/2.89[a]	1.74/1.8	0.863/0.579	18.490/9.375
Pouring method	4.31/5.78	1.97/2.22	0.759/0.883	19.091/33.774
Airless method	6.97/9.78	3.32/2.5	0.705/1.267	53.966/94.109

[a]Above the line shows the values for the viscosity of the ink = 0.001 Pa s; below the line is for viscosity = 0.00026 Pa s.

Table 5.5 Financial losses of the enterprise depending on reproducibility of process of coloring

Index of reproducibility C_{pk}	0.579	0.705	0.759	0.863	0.883	1.267
Losses $L(y)$ (ruble/100 m^2)	45,734.2	263,258.9	93,130.8	90,200.1	164,756.1	459,080

expenses $L(y)$. Thus at an index of reproducibility of C_{pk} of = 0.579 losses are minimal and make $L(y) = 45.734$ ruble/100 m^2, while at $C_{pk} = 1.267 - 459.080$ ruble/100 m^2. A bad adjustment of the process completely destroys all potential advantages of reproducibility improvement. If the process is not adjusted based on the target value y_o it is impossible to judge the process' efficiency based on the index of reproducibility C_{pk}, considering that $C_{pk} > 1$ indicates the process is effective. It is necessary to consider the additional expenses connected to the loss of quality of production. Of course, the closer the average process to the target value y_o and the less spread of quality indicators, the lower the losses of the manufacturer.

The abovementioned results show the importance of exact adjustment of the coloring process of building products and designs.

Thus the quality of the production process is determined by the financial losses due to deviations of quality indicators from the target values. To optimize the process to be applied technological methods, organization of the process, etc.

5.2 Use of statistical methods to assess the cracking of paint and varnish coatings

The most common types of damage are peeling and cracking. The conditions of adhesion and cohesion damage can be written as:

- Adhesion failure

$$R_a < R_k \tag{5.6}$$

- Cohesive failure

$$R_k < R_a \tag{5.7}$$

where R_a and R_k are, respectively, the adhesive and cohesive strength of the coatings.

Knowing the distribution laws (probability density) $f(R_a)$ and $f(R_k)$, it is possible to determine the dependence of the adhesion failure probability P_a on the observed

R_k value and, similarly, the dependence of the cohesion failure probability P_k on the observed R_a value as follows:

$$P_a(R_k) = \int_{-\infty}^{R_k} f(R_a)dR_a \tag{5.8}$$

$$P_k(R_a) = \int_{-\infty}^{R_a} f(R_k)dR_k \tag{5.9}$$

The solution of this task includes the following stages:

- collection of statistical material, characterizing the spread of the limiting values of R_k and R_a of the test coating;
- construction of histograms, determination of mean values, standard deviations, testing of hypotheses on compliance with standard deviations; and
- construction of the functions $P_a(R_k)$ and $P_k(R_a)$.

Consider the probabilities of cohesive and adhesive failure of polyvinyl acetate-cement (PVAC) and polymer-calcareous coatings on a cement substrate.

Tensile adhesive strength tests (R_a) and tensile cohesive strength (R_k) were subjected to 20 coating samples of each type. The test results in the form of variational series are given in Table 5.6.

The statistical distributions for R_a and R_k of the coatings under study are given in Tables 5.7–5.10.

The histograms of the distribution of the relative frequencies of the adhesion and cohesion strength of the PVAC coating are shown in Fig. 5.2, and the histograms of the distribution of the relative frequencies of the adhesion and cohesion strength of polymer-calcareous coating are shown in Fig. 5.3.

These histograms allow us to put forward a hypothesis about normal distributions for all considered sets. The results of the hypothesis testing by the Pearson criterion χ^2 (significance level $\alpha = 0.05$, number of degrees of freedom $k = s - 3$, where s is the number of intervals) are given in Table 5.11.

The results obtained show good agreement between the empirical distribution law and the normal distribution. Thus the distribution functions will have the form:

$$f(R_a) = \frac{1}{\sqrt{2\pi}\sigma_{R_a}} \times e^{-\frac{1}{2}\frac{(R_a-\overline{R}_a)^2}{\sigma^2_{R_a}}} \tag{5.10}$$

$$f(R_k) = \frac{1}{\sqrt{2\pi}\sigma_{R_k}} \times e^{-\frac{1}{2}\frac{(R_k-\overline{R}_k)^2}{\sigma^2_{R_k}}} \tag{5.11}$$

The graphs of the dependencies of the probabilities of the adhesive $R_a(R_k)$ and the cohesive $P_k(R_a)$ of the coating destruction in accordance with formulas (5.10) and (5.11) are shown in Figs. 5.4 and 5.5.

Table 5.6 Presented in the form of a variational series

No.	PVAC coating		No.	Polymer-calcareous coating	
	R_k (MPa)	R_a (MPa)		R_k (MPa)	R_a (MPa)
1	1.4	1.0	1	1.1	0.9
2	1.7	1.3	2	1.3	1.1
3	1.8	1.4	3	1.3	1.2
4	1.8	1.4	4	1.5	1.3
5	1.9	1.5	5	1.5	1.3
6	1.9	1.5	6	1.5	1.4
7	2.0	1.6	7	1.6	1.4
8	2.0	1.6	8	1.6	1.5
9	2.0	1.7	9	1.7	1.5
10	2.1	1.7	10	1.7	1.5
11	2.1	1.7	11	1.7	1.5
12	2.2	1.8	12	1.8	1.6
13	2.2	1.8	13	1.8	1.6
14	2.2	1.8	14	1.8	1.6
15	2.3	1.8	15	1.8	1.7
16	2.3	1.9	16	1.9	1.7
17	2.3	1.9	17	1.9	1.8
18	2.5	2.1	18	2.1	1.9
19	2.5	2.1	19	2.1	2.0
20	2.8	2.4	20	2.3	2.1

Table 5.7 Distribution of R_a PVAC coating

Value (MPa)	1.0	1.3	1.4	1.5	1.6	1.7	1.8	1.9	2.1	2.4
Frequency, n_i	1	1	2	2	2	3	4	2	2	1

Mean: $\overline{R_a} = 1.7$ MPa; standard deviations: $\sigma_{R_a} = 0.313$ MPa.

Table 5.8 Distribution R_k of PVAC coating

Value (MPa)	1.4	1.7	1.8	1.9	2.0	2.1	2.2	2.3	2.5	2.8
Frequency, n_i	1	1	2	2	3	3	3	3	2	1

Mean: $\overline{R_k} = 2.1$ MPa; standard deviations $\sigma_{R_k} = 0.308$ MPa.

Table 5.9 Distribution of R_a of the polymer-calcareous coating

Value (MPa)	0.9	1.1	1.2	1.3	1.4	1.5	1.6	1.7	1.8	1.9	2.0	2.1
Frequency, n_i	1	1	1	2	2	4	3	2	1	1	1	1

Mean: $\overline{R_a} = 1.53$ MPa; standard deviations: $\sigma_{R_a} = 0.296$ MPa.

Table 5.10 Distribution of R_k of the polymer-calcareous coating

Value (MPa)	1.1	1.3	1.5	1.6	1.7	1.8	1.9	2.1	2.3
Frequency, n_i	1	2	3	2	3	4	2	2	1

Mean: $\overline{R_k} = 1.7$ MPa; standard deviations: $\sigma_{R_k} = 0.292$ MPa.

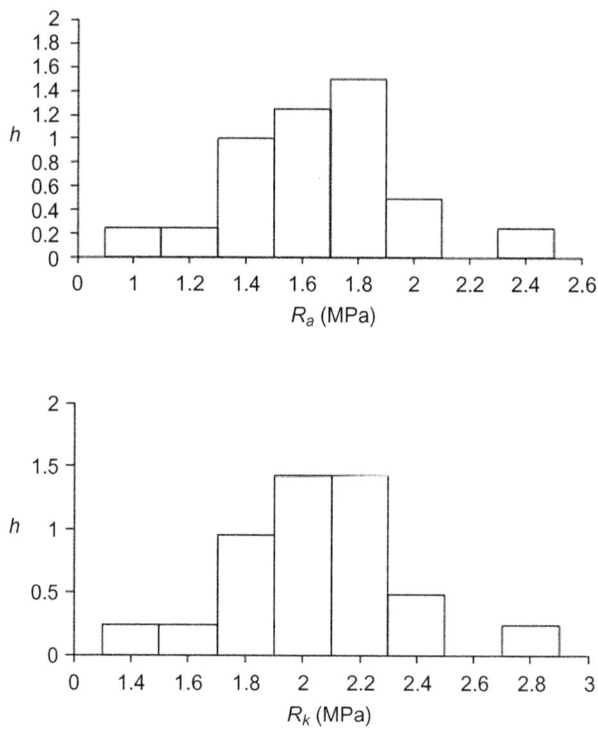

Figure 5.2 Histograms of the distribution of relative frequencies—adhesion and cohesion strength of PVAC coatings.

Analysis of the obtained probabilities allows us to show that for PVAC coatings in approximately 85% of the cases $\{P_a(\overline{R_k} = 2.1 \text{ MPa}) = 0.85\}$, adhesion "renouncement" will be observed and in 15% of cases cohesive "renouncement" (cracking) will be seen. It is clear that research aimed primarily at increasing the adhesion strength (analysis of the roughness and porosity of the substrate, compatibility of the coating and substrate, characteristics of the contact layer) is needed.

For polymer-calcareous coating adhesion, "renouncement" can be approximately predicted in 70% of cases $\{P_a(\overline{R_k} = 1.7 \text{ MPa}) = 0.7\}$, and for cohesive, in 30% of cases.

The obtained results can also be used to establish critical levels of adhesion strength during the acceptance inspection of coatings at sites. Thus for PVAC coverage, a critical level of adhesion strength can be established for an acceptance test

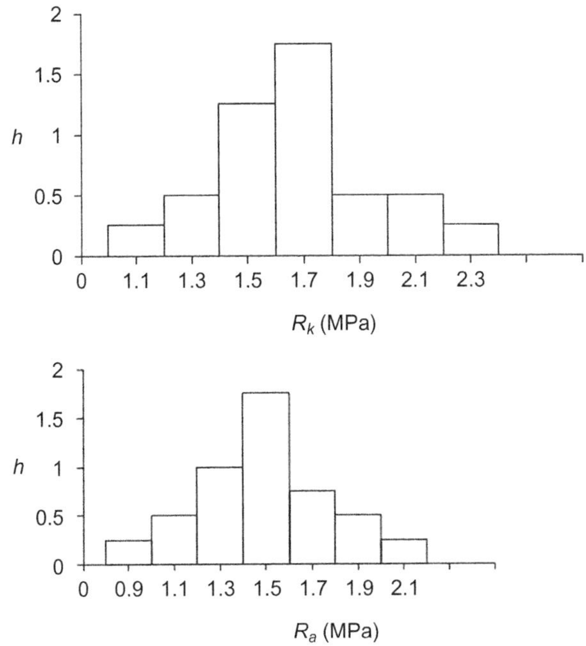

Figure 5.3 Histograms of the distribution of the relative frequencies of the adhesive and cohesive strength of the polymer-calcareous coating.

Table 5.11 Results of hypothesis testing by the Pearson criterion $\chi 2$

Coating	Function	$\chi^2_{набл}$	$\chi^2_{крит}$
PVAC	$f(R_a)$	3.62	11.1
	$f(R_k)$	3.84	11.1
Polymer-calcareous	$f(R_a)$	0.47	9.5
	$f(R_k)$	3.21	9.5

with an alternative feature: $R_a = 1.2$ MPa, at which the probability $P_k = 0.003$ (practically eliminated) makes the control more objective.

Let us consider the probability of cohesive destruction of PVAC of the coating $P(R_k < R_a)$ provided that the thickness of the coating takes random values equal to h.

In order to determine the probability of cracking of the coating under study, tests were carried out of stained samples with different coating thicknesses (20 samples for each thickness) to determine the tensile strength. The data of tests in the form of variational series are given in Table 5.12.

As is known, the change in the cohesive strength of a coating as a function of thickness is most accurately described in accordance with the Weibull equation, namely:

$$\overline{R_k} = K \times h^{-\frac{1}{n}} \tag{5.12}$$

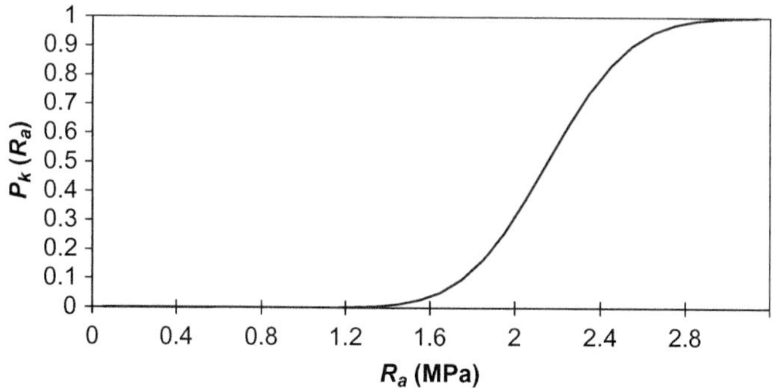

Figure 5.4 The probability of adhesion $R_a(R_k)$ and cohesive $P_k(R_a)$ destruction of polymer-calcareous coating.

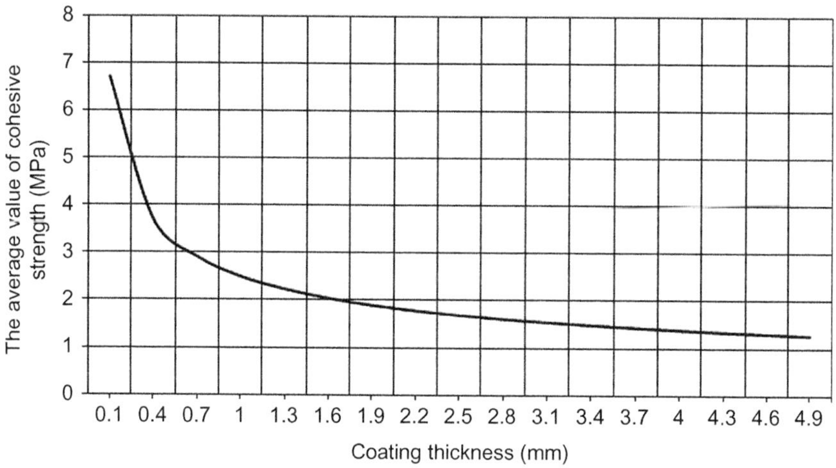

Figure 5.5 Dependence of cohesive strength of coatings on their thickness.

where h is the thickness of the coating and K and n are coefficients depending on the type of coating.

The average quadratic deviation σ_{R_k} is also, as $\overline{R_k}$ it will be, depending on the observed thickness of the coating. According to the data (Table 5.12), the dependence σ_{R_k} on h can be represented in the form:

$$\sigma_{R_k} = A - B \times e^{-C \times h^d} \tag{5.13}$$

where h is the thickness of the coating and A, B, C, d are the coefficients that depend on the type of coating.

Having carried out calculations using the method of least squares, we have established the dependences $\overline{R_k(h)}$ and σ_{R_k} that have the form:

$$\overline{R_k} = 2.5 \times h^{-\frac{1}{2.33}} \tag{5.14}$$

Table 5.12 Changes in the cohesive strength of the coatings as a function of their thickness

No.	Cohesive strength R_k (MPa)				
	Coating thickness h (mm)				
	0.5	**1.0**	**1.5**	**2.0**	**2.5**
1	2.1	1.9	1.4	1.4	1.2
2	2.2	2.1	1.7	1.5	1.3
3	2.2	2.1	1.8	1.6	1.4
4	2.3	2.1	1.8	1.7	1.5
5	2.3	2.2	1.9	1.7	1.6
6	2.4	2.2	1.9	1.8	1.6
7	2.4	2.2	2.0	1.8	1.6
8	2.5	2.3	2.0	1.8	1.7
9	2.6	2.4	2.0	1.9	1.7
10	2.7	2.5	2.1	1.9	1.7
11	2.7	2.5	2.1	1.9	1.7
12	2.8	2.6	2.2	1.9	1.7
13	2.9	2.7	2.2	2.0	1.8
14	2.9	2.7	2.2	2.0	1.8
15	3.0	2.8	2.3	2.0	1.8
16	3.1	2.8	2.3	2.1	1.9
17	3.1	2.9	2.3	2.1	1.9
18	3.2	2.9	2.5	2.3	2.1
19	3.2	3.0	2.5	2.4	2.2
20	3.4	3.1	2.8	2.6	2.3
Mean $\overline{R_k}$	2.7	2.5	2.1	1.9	1.7
Standard deviations σ_{R_k}	0.391	0.352	0.311	0.291	0.275

$$\sigma_{R_k} = 0.4 - 0.15 \times e^{-h^{-1.8}} \tag{5.15}$$

Graphically, the dependences (5.14) are shown in Fig. 5.6.

The value $\overline{R_k}$ is random and obeys the normal distribution law (as confirmed in Eq. 5.1). In accordance with the formulas of the normal distribution, the distribution law R_k will have the form:

$$f(R_k) = \frac{1}{\sqrt{2\pi} \times (A - B \times e^{-C \times h^d})} \times e^{-\frac{1}{2}\frac{(R_k - K \times h^{-\frac{1}{n}})^2}{(A - B \times e^{-C \times h^d})^2}} \tag{5.16}$$

The probability of cohesive failure (cracking) is calculated by the formula

$$P(R_k < R_a) = \int_{-\infty}^{R_a} f(R_k) dR_k \tag{5.17}$$

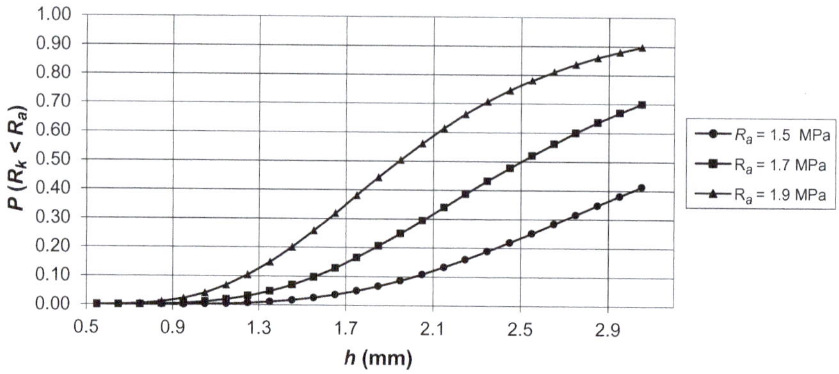

Figure 5.6 The dependence of the probability of cracking on the thickness h of the coating.

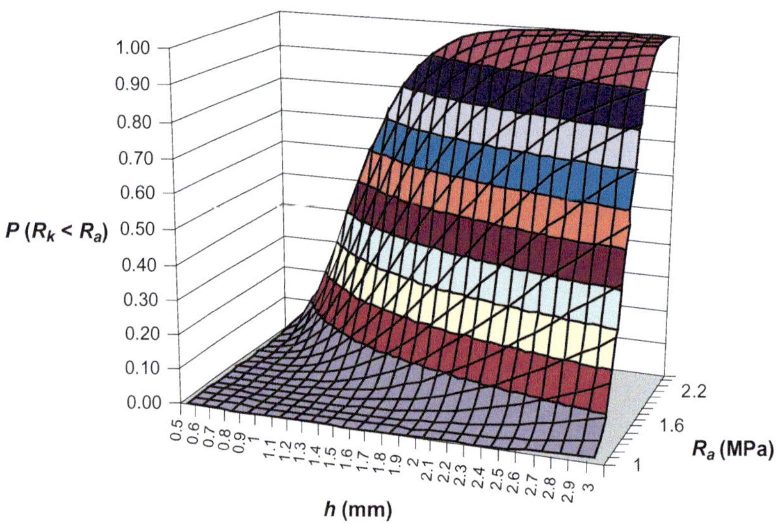

Figure 5.7 Dependence of the probability of cohesive failure of coating from thickness h and adhesion strength R_a.

Thus the probability of cracking of any part of the coating surface will be determined by the current values of the adhesion strength and the thickness of coating.

In accordance with Eq. (5.13), the probability of cohesive destruction of $P(R_k < R_a)$ as a function of h for different values of R_a (most probable for the coating under consideration) will be described by the curves shown in Fig. 5.7.

In general, the probability $P(R_k < R_a)$ as a function of h and R_a can be described by the surface shown in Fig. 5.7.

Applying the constructed surface (or its analytic form 5.17), we give some values of the cohesive failure probabilities for various h and R_a (limiting extreme and average). The values are given in Table 5.13. Волее полная таблица,

Table 5.13 Some values of the probabilities of cohesive failure

Adhesion strength value (MPa)		*h* (mm)	
		0.5	**3**
R_a (MPa)	1.0	1.05×10^{-9}	0.019
	1.7	0.000012	0.6999
	2.5	0.014	0.9999

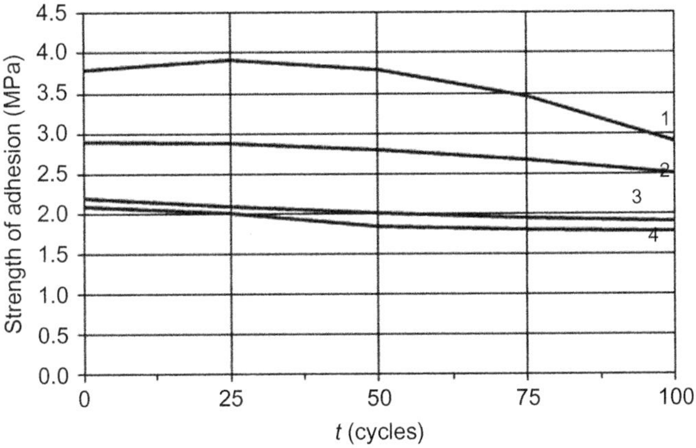

Figure 5.8 Change in adhesion strength $M_{R_a}(t)$in the process of humidification-drying PVAC coating on substrates with different porosity:
1—porosity $P = 1.9\%$
2—porosity $P = 2.7\%$
3—porosity $P = 3.4\%$
4—porosity $P = 5.9\%$

содержащая значения рассматриваемых вероятностей приведена в приложении 1.

Elimination of the cohesive nature of fracture implies higher values of cohesive strength and creates prerequisites in combination with materials science factors for creating crack-resistant coatings of optimal thickness (Fig. 5.8).

The applied approach can be applied in the development of quality control of finishing on an alternative based.

Any characteristic of the coating can be represented as:

$$R_i = f(a_1^i, a_2^i, \ldots, a_n^i, t) \tag{5.18}$$

where R_i is the coverage characteristic; $R_i = f(a_1^i, a_2^i, \ldots, a_n^i, t)$ are the indicators that determine the characteristics of the coating; and t is the time indicator (duration of thermal aging, UV irradiation, number of humidification cycles, etc.).

An investigation of the behavior of the R_i characteristics of a covering using mathematical statistics methods consists of comparing their estimates in the course of a change in the index t. Such a comparison will allow us to determine the prevalence of the probability of a particular type of destruction, to predict (with a certain accuracy) the quality of the coatings and to identify the characteristics most affected by the time index. It should be noted that the objectivity of the conclusions based on the results of the studies will be strengthened with the increase in sample volumes and a decrease in the time intervals between measurements.

Let us consider an example of a definition of a statistical estimation of probability of adhesive destruction PVAC of the coverings formed on cement substrates with various porosity, during carrying out of cycles "humidifying-drying." The condition for adhesive failure was determined by the expression (5.18).

The value of R_a in the example under consideration can be written as:

$$R_a = f(P, t) \tag{5.19}$$

where P is the porosity of the substrate.

The cohesive strength due to porosity depends weakly, and in the problem under consideration this dependence can be neglected, and therefore we will adopt (Table 5.14).

$$R_k = f(t) \tag{5.20}$$

At any moment in time, the quantities R_a and R_k will be random and distributed, in general, according to normal laws, which will be determined by the following mathematical expectations and standard deviations: $M_{R_a}(\Pi, t)$, $\sigma_{R_a}(\Pi, t), M_{R_k}(t), \sigma_{R_k}(t)$.

The probability of adhesion failure of the coating will be determined as the value of the distribution function of the two-dimensional random variable $F(R_a, R_k)$, determined, subject to condition (5.17), as

$$P(R_a < R_k) = \int_0^{\infty} \int_{R_a}^{\infty} f(R_k) \times f(R_a) dR_k dR_a \tag{5.21}$$

where

$$f(R_k) = \frac{1}{\sqrt{2\pi}\sigma_{R_k}(t)} \times e^{-\frac{1}{2}\frac{(R_k - M_{R_k}(t))^2}{\sigma^2_{R_k}(t)}} \tag{5.22}$$

$$f(R_a) = \frac{1}{\sqrt{2\pi}\sigma_{R_a}(\Pi, t)} \times e^{-\frac{1}{2}\frac{(R_a - M_{R_a}(\Pi, t))^2}{\sigma^2_{R_a}(\Pi, t)}} \tag{5.23}$$

Table 5.14 Adhesive and cohesive strength values coatings at the initial time $t = 0$

No.	Adhesive strength R_a (MPa)				R_k (MPa)
	Porosity of substrate P (%)				
	1.9	**2.7**	**3.4**	**5.9**	
1	3.3	2.5	1.9	1.8	1.4
2	3.3	2.6	1.9	1.9	1.7
3	3.4	2.6	2.0	1.9	1.8
4	3.4	2.7	2.0	2.0	1.8
5	3.5	2.8	2.0	2.0	1.9
6	3.5	2.8	2.1	2.0	1.9
7	3.7	2.9	2.1	2.0	2.0
8	3.7	2.9	2.2	2.0	2.0
9	3.7	2.9	2.2	2.1	2.0
10	3.8	2.9	2.2	2.1	2.1
11	3.8	2.9	2.2	2.1	2.1
12	3.8	3.0	2.2	2.1	2.2
13	3.9	3.0	2.2	2.1	2.2
14	3.9	3.1	2.3	2.1	2.2
15	4.0	3.1	2.3	2.1	2.3
16	4.1	3.1	2.4	2.2	2.3
17	4.1	3.2	2.4	2.2	2.3
18	4.1	3.2	2.4	2.3	2.5
19	4.2	3.3	2.5	2.4	2.5
20	4.3	3.4	2.5	2.5	2.8
Mean, $M_{R_a}(\Pi, t)$, $M_{R_k}(t)$	3.8	2.9	2.2	2.1	2.1
Standard deviations, $\sigma_{R_a}(\Pi, t)$, $\sigma_{R_k}(t)$	0.304	0.239	0.184	0.167	0.310

Thus the probability $P(R_a < R_k)$ will be a function that depends on the time index and porosity of the substrate.

In order to determine the peeling probabilities of the coating under study, tests were carried out of stained samples with different porosity of the substrate (20 samples for each porosity) to determine the change in adhesion strength with successive humidification and drying of the coatings. Simultaneously, tests were conducted to determine the tensile strength. The data of the tests in the form of variational series are given in Tables 5.15 and 5.16.

The graphically investigated dependences are shown in Figs. 5.9–5.12.

Carrying out the calculations according to expression (5.23), we obtain the graphs of the change in the probability of peeling PVAC coatings on substrates with different porosity in the process of humidification-drying ($P(R_a < R_k)$ (Fig. 5.12).

Table 5.15 Adhesive and cohesive strength values of coatings after 50 cycles of humidification-drying ($t = 50$)

No.	Adhesive strength R_a (MPa)				R_k (MPa)
	Porosity of substrate P (%)				
	1.9	**2.7**	**3.4**	**5.9**	
1	2	3	4	5	6
1	3.3	2.4	1.6	1.5	1.1
2	3.3	2.5	1.7	1.6	1.2
3	3.4	2.5	1.7	1.7	1.2
4	3.4	2.5	1.7	1.7	1.3
5	3.5	2.7	1.8	1.7	1.3
1	2	3	4	5	6
6	3.5	2.7	1.8	1.8	1.3
7	3.7	2.8	1.9	1.8	1.3
8	3.7	2.8	2.0	1.8	1.3
9	3.7	2.8	2.0	1.8	1.4
1	2	3	4	5	6
10	3.8	2.8	2.0	1.8	1.4
11	3.8	2.8	2.0	1.8	1.4
12	3.8	2.9	2.0	1.8	1.4
13	3.9	2.9	2.0	1.9	1.4
14	3.9	3.0	2.1	1.9	1.4
15	4.0	3.0	2.2	1.9	1.4
16	4.1	3.0	2.2	1.9	1.5
17	4.1	3.1	2.2	2.0	1.5
18	4.2	3.2	2.3	2.1	1.6
19	4.2	3.2	2.3	2.1	1.7
20	4.3	3.3	2.4	2.1	1.8
Mean, $M_{R_a}(\Pi, t)$, $M_{R_k}(t)$	3.8	2.8	2.0	1.84	1.4
Standard deviations, $\sigma_{R_a}(\Pi, t)$, $\sigma_{R_k}(t)$	0.311	0.252	0.228	0.160	0.167

The obtained dependences make it possible to draw the following conclusions:

- At the initial moment of time, a decrease in the porosity of the substrate from 3.4% to 1.9% leads to a sharp decrease (by four orders of magnitude) in the probability of adhesive failure, and an increase in porosity from 3.4% to 5.9% does not practically change the probability of adhesive failure.
- For coatings on all substrates there is a decrease in the probability of adhesive failure with increasing number of cycles. This indicates a stronger effect of moisture on the decrease in cohesive strength of coatings. But it should be noted that on substrates with a porosity of 1.9% and 2.7% adhesion failure is higher than on substrates with a porosity of 3.4% and 5.9%.
- The processes of changing the strength characteristics of coatings on substrates with a porosity of 3.4% and 5.9% can be considered identical, which allows us to propose a

Table 5.16 Adhesive and cohesive strength values of coatings after 100 cycles of humidification-drying ($t = 100$)

No.	Adhesive strength R_a (MPa)				R_k (MPa)
	Porosity of substrate P (%)				
	1.9	**2.7**	**3.4**	**5.9**	
1	2	3	4	5	6
1	2.5	2.1	1.6	1.5	0.7
2	2.6	2.2	1.6	1.6	0.7
3	2.6	2.2	1.7	1.6	0.7
4	2.7	2.2	1.7	1.7	0.8
5	2.8	2.3	1.7	1.7	0.8
6	2.8	2.3	1.7	1.7	0.8
7	2.9	2.4	1.8	1.7	0.8
8	2.9	2.5	1.9	1.7	0.8
9	2.9	2.5	1.9	1.7	0.8
10	2.9	2.5	1.9	1.7	0.9
11	2.9	2.5	1.9	1.8	0.9
12	3.0	2.5	1.9	1.8	0.9
13	3.0	2.5	1.9	1.8	0.9
14	3.1	2.6	2.0	1.8	0.9
15	3.1	2.7	2.1	1.9	0.9
1	2	3	4	5	6
16	3.1	2.7	2.1	1.9	0.9
17	3.2	2.7	2.1	1.9	1.0
18	3.2	2.8	2.1	2.0	1.0
19	3.3	2.8	2.2	2.1	1.0
20	3.4	2.9	2.3	2.1	1.1
Mean, $M_{R_a}(\Pi, t)$, $M_{R_k}(t)$	2.9	2.5	1.9	1.79	0.9
Standard deviations, $\sigma_{R_a}(\Pi, t)$, $\sigma_{R_k}(t)$	0.239	0.228	0.201	0.160	0.109

hypothesis about the existence of such a porosity of the substrate for PVAC coatings (e.g., 3.4%), which can be considered boundary, separating the areas of high and weak sensitivity of properties to the effects of moisture.

The above methodology and example can be used to develop requirements for the protective properties of paint and varnish coatings for various surfaces and specific operating conditions, as well as to solve the problems of predicting the protective properties of coatings.

As already noted, cracking of coatings occurs when internal tensile stresses reach the cohesive strength of the coating material. The increase in internal stresses and the reduction of cohesive strength occurs, primarily, due to the impact on the coating of various climatic factors during operation. Reduction of the strength characteristics of coatings under operating conditions should be

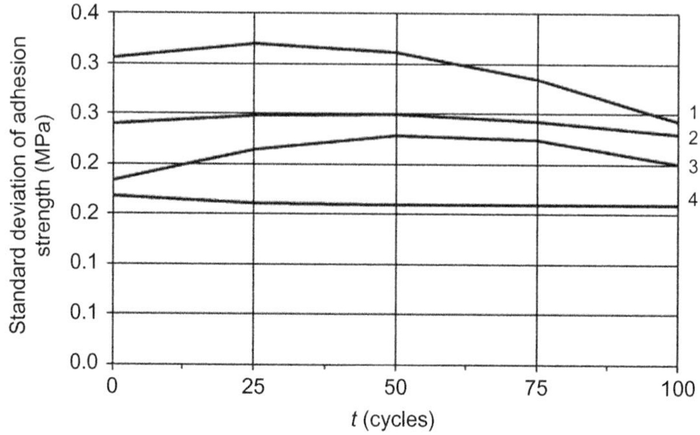

Figure 5.9 Change in standard deviations strength of adhesion $\sigma_{R_a}(t)$ in the process of humidification-drying PVAC coating on substrates with different porosity:
1—porosity $P = 1.9\%$
2—porosity $P = 2.7\%$
3—porosity $P = 3.4\%$
4—porosity $P = 5.9\%$

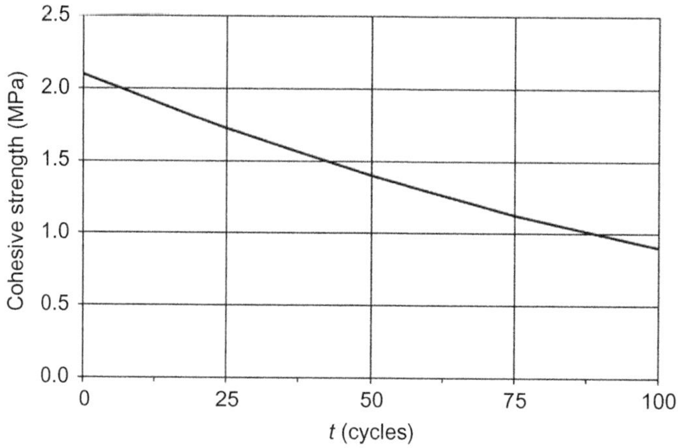

Figure 5.10 Change in cohesive strength during humidification-drying of PVAC coatings.

presented as random processes, the mathematical expectation and standard deviations of which is represented as a function of some time indicator (operation period, thermal aging, UV exposure, humidification, etc.). Comparison of different characteristics of the coating at certain points in time makes it possible to estimate the probabilities of various kinds of destruction, the study of which in laboratory conditions should ensure an objective prediction of the quality and

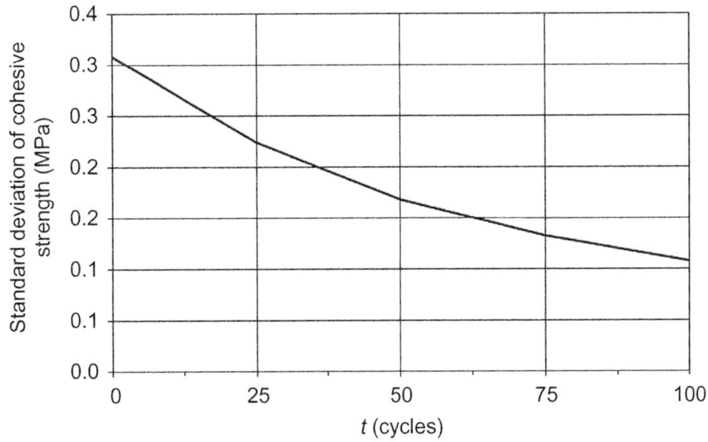

Figure 5.11 Change in standard deviations cohesive strength in the process of humidification-drying PVAC coatings.

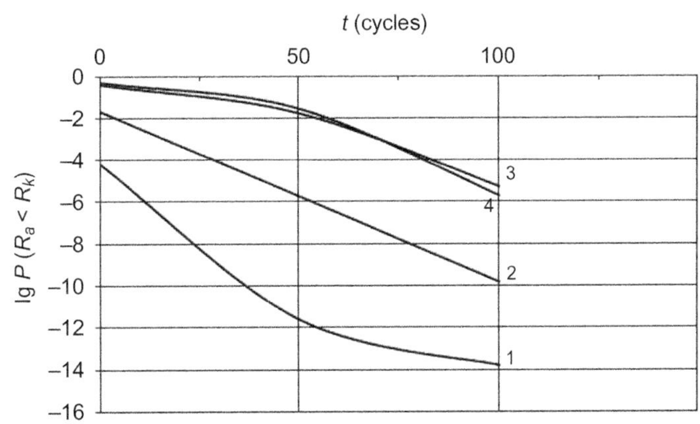

Figure 5.12 The change in the probability of adhesive failure in the dependence from the number of cycles of humidification-drying:
1—substrate with porosity $P = 1.9\%$
2—substrate with porosity $P = 2.7\%$
3—substrate with porosity $P = 3.4\%$
4—substrate with porosity $P = 5.9\%$

reliability of the coatings and, accordingly, the fulfillment of the requirements and expectations of the consumer.

The change in $R_{\kappa o z}$ is a random process, at each time of which the distribution of the random variable obeys the normal law with mathematical expectation $M_{R_k}(t)$ and

Table 5.17 Values of cohesive strength and internal stresses in PVAC coatings in the process of thermal aging

No.	Cohesive strength R_{kog} (MPa)			
	Time t (h)			
	0	**50**	**100**	**150**
1	2	3	4	5
1	1.6	4.5	3.7	3.2
2	1.6	4.5	3.7	3.3
3	1.7	4.5	3.8	3.4
4	1.8	4.6	3.8	3.4
5	1.8	4.6	3.9	3.5
6	1.9	4.7	3.9	3.6
7	1.9	4.8	4.1	3.6
8	1.9	4.9	4.3	3.7
9	1.9	5.2	4.5	3.9
10	2.0	5.4	4.6	4.1
11	2.1	5.5	4.6	4.1
12	2.1	5.7	4.8	4.3
13	2.2	5.9	4.9	4.4
14	2.2	6.1	5.1	4.4
15	2.3	6.2	5.2	4.5
1	2	3	4	5
16	2.4	6.3	5.3	4.7
17	2.4	6.3	5.4	4.8
18	2.5	6.6	5.5	4.9
19	2.5	6.7	5.6	5.0
20	2.6	6.8	5.7	5.1
Mean, $M_{R_k}(t)$	2.1	5.49	4.62	4.1
Standard deviations, $s_{R_k}(t)$	0.308	0.823	0.693	0.615
σ (t) (MPa)	0.06	0.142	0.162	0.162

standard deviations $S_{R_k}(t)$. The quantity σ is considered as a function of time $\sigma(t)$. Then probability $P(R_{\kappa o \varepsilon} < \sigma)$ will also be a function of time and will be defined as:

$$P(R_{\kappa o \varepsilon} < \sigma) = \int_{-\infty}^{\sigma(t)} \frac{1}{\sqrt{2\pi} \times s_{R_k}(t)} \times e^{-\frac{1}{2}\frac{(R_k - M_{R_k}(t))^2}{s_{R_k}^2(t)}} \quad (5.24)$$

In order to determine the probability of cracking of PVAC and polymer-based coatings, 20 samples of stained samples were tested for function $M_{R_k}(t)$, standard deviations determination and $\sigma(t)$ during thermal aging (holding the samples at $+50°C$) and during UV irradiation.

The data of tests in the form of variational series are given in Tables 5.17–5.20.

Table 5.18 Values of cohesive strength and internal stresses in polymer-calcareous coatings in the process of heat aging

No.	Cohesive strength R_{kog} (MPa)			
	Time t (h)			
	0	**50**	**100**	**150**
1	1.4	2.1	1.6	1.5
2	1.4	2.2	1.6	1.5
3	1.5	2.2	1.7	1.6
4	1.6	2.2	1.8	1.7
5	1.7	2.4	1.9	1.8
6	1.7	2.5	1.9	1.9
7	1.7	2.6	1.9	1.9
8	1.7	2.7	2.0	2.0
9	1.8	2.7	2.0	2.0
10	1.9	2.7	2.0	2.0
11	1.9	2.7	2.1	2.1
12	1.9	2.8	2.2	2.2
13	1.9	2.8	2.2	2.2
14	2.0	2.8	2.3	2.2
15	2.0	3.0	2.3	2.2
16	2.1	3.1	2.4	2.3
17	2.2	3.2	2.5	2.3
18	2.3	3.3	2.5	2.4
19	2.3	3.4	2.6	2.5
20	2.4	3.5	2.6	2.5
Mean, $M_{R_k}(t)$	1.87	2.75	2.11	2.04
Standard deviations, $s_{R_k}(t)$	0.292	0.412	0.317	0.306
σ (t) (MPa)	0.025	0.03	0.07	0.07

Approximating functions that give the fewest errors when applying the least squares method to obtain analytical dependencies of the processes under study are given in Table 5.20.

The graphically investigated dependences are shown in Figs. 5.13 and 5.14.

Analysis of graphical dependencies allows for the following to be noted:

- For PVAC coatings at the initial moment of time, at thermal aging and UV irradiation, a sharp increase in cohesive strength is observed, which also sharply decreases after >100 hours of testing. With UV irradiation, the decrease is less dramatic than in the case of thermal aging. The growth of internal stresses in the samples of coatings occurs relatively uniformly, and under thermal aging, the internal stresses are 1.5–2 times higher in absolute magnitude than with UV irradiation.

Table 5.19 Values of cohesive strength and internal stresses in PVAC coatings at UV irradiation

No.	Cohesive strength R_{kog} (MPa)			
	Time t (h)			
	0	50	100	150
1	1.6	3.7	3.5	3.5
2	1.6	3.8	3.6	3.6
3	1.7	3.8	3.6	3.6
4	1.8	3.9	3.7	3.7
5	1.8	3.9	3.7	3.7
6	1.9	4.0	3.8	3.8
7	1.9	4.3	3.8	3.8
8	1.9	4.4	4.1	4.1
9	1.9	4.5	4.3	4.3
10	2.0	4.6	4.4	4.4
11	2.1	4.6	4.4	4.4
12	2.1	4.6	4.4	4.4
13	2.2	4.7	4.5	4.5
14	2.2	4.9	4.8	4.6
15	2.3	5.2	4.9	4.7
16	2.4	5.3	5.1	4.9
17	2.4	5.4	5.2	5.1
18	2.5	5.5	5.3	5.2
19	2.5	5.6	5.4	5.4
20	2.6	5.6	5.4	5.4
Mean $M_{R_k}(t)$	2.1	4.62	4.4	4.36
Standard deviations $s_{R_k}(t)$	0.308	0.647	0.660	0.621
$\sigma(t)$ (MPa)	0.05	0.065	0.093	0.097

- For polymer-calcareous coatings, a slight increase in cohesive strength is observed at the initial moment of time, but not as sharp as for PVAC coatings, which after 100 hours of testing starts to decrease smoothly. In absolute value, the cohesive strength is 1.5–2 times less than for PVAC coatings. Internal stresses in polymer-calcareous coatings during thermal aging are 1.5–2 times lower throughout the test period, and under UV irradiation after 150 hours of testing, internal stresses in polymer-calcareous coatings exceed internal stresses in coatings of PVAC and increase more intensively [16–18].

Carrying out the calculations according to the expression (5.24), we obtain the graphs of the change in the probability of cracking of PVAC and polymer-calcareous coatings ($P(R_{\kappa o z} < \sigma)$) under thermal aging and UV irradiation (Figs. 5.15 and 5.16).

Table 5.20 Values of cohesive strength and internal stresses in polymer-calcareous coatings at UV irradiation

No.	Cohesive strength R_{kog} (MPa)			
	Time t (h)			
	0	**50**	**100**	**150**
1	2	3	4	5
1	1.4	1.7	1.5	1.3
2	1.4	1.8	1.6	1.3
3	1.5	1.8	1.7	1.4
4	1.6	1.9	1.8	1.5
1	2	3	4	5
5	1.7	1.9	1.8	1.5
6	1.7	2.0	1.9	1.5
7	1.7	2.0	1.9	1.5
8	1.7	2.1	2.0	1.5
9	1.8	2.2	2.0	1.6
10	1.9	2.2	2.1	1.7
11	1.9	2.2	2.1	1.7
12	1.9	2.3	2.2	1.7
13	1.9	2.3	2.3	1.7
14	2.0	2.3	2.3	1.8
15	2.0	2.5	2.3	1.8
16	2.1	2.6	2.4	1.9
17	2.2	2.6	2.4	2.0
18	2.3	2.7	2.5	2.0
19	2.3	2.7	2.6	2.1
20	2.4	2.8	2.6	2.1
Mean $M_{R_k}(t)$	1.87	2.23	2.1	1.68
Standard deviations $s_{R_k}(t)$	0.292	0.334	0.325	0.248
$\sigma(t)$, MPa	0.025	0.07	0.090	0.099

The analysis of the obtained dependences in combination with the results of the studies shown in Figs. 5.15 and 5.16 allows us to draw the following conclusions:

- With thermal aging, the probability of cracking of PVAC coatings is an order of magnitude higher than that of polymer-calcareous coatings, despite the higher average value of cohesive strength. This is due to a more intense increase in internal stresses and a large spread of observed values near the mean. This observation allows us to conclude that the stability of polymer-calcareous coatings is higher for prolonged exposure to elevated temperatures with respect to PVAC coatings.
- With UV irradiation at the initial moment of time, the probability of cracking of polymer-calcareous coatings begins to increase rapidly, while the probability of cracking of PVAC coatings begins to increase after 150 hours of testing. Thus PVAC coatings are more resistant to UV irradiation (as opposed to thermal aging) with respect to polymer-calcareous coatings.

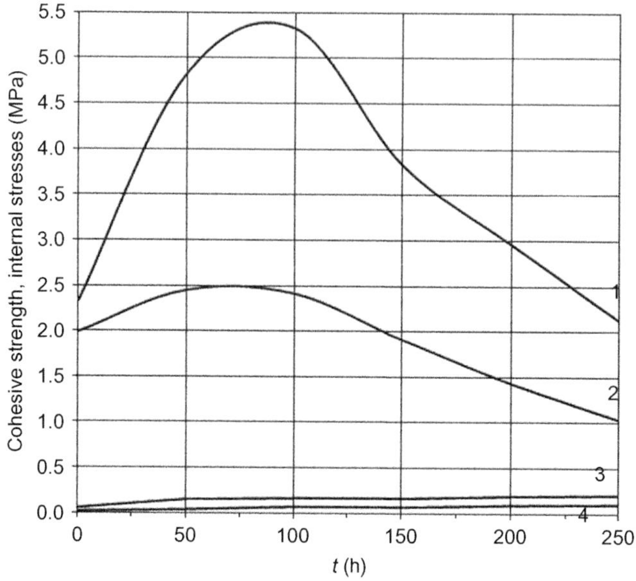

Figure 5.13 Changes in cohesive strength $M_{R_k}(t)$ and internal stresses during thermal aging of PVAC and polymer-calcareous coatings:
1—change $M_{R_k}(t)$ of PVAC coatings
2—change $M_{R_k}(t)$ of polymer-calcareous coatings
3—change of $\sigma(t)$ PVAC coatings
4—change of $\sigma(t)$ polymer-calcareous coatings

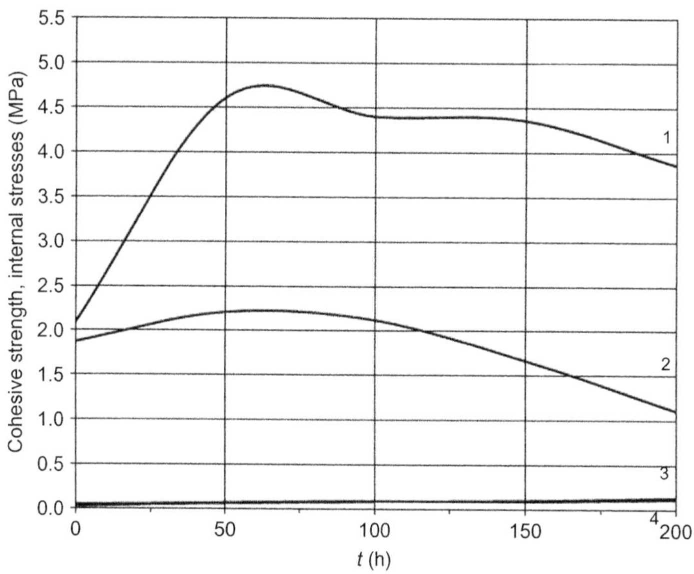

Figure 5.14 Change in cohesive strength $M_{R_k}(t)$ and internal stresses during UV irradiation of PVAC and polymer-calcareous coatings:
1—change $M_{R_k}(t)$ of PVAC coatings
2—change $M_{R_k}(t)$ of polymer-calcareous coatings
3—change of $\sigma(t)$ PVAC coatings
4—change in $\sigma(t)$ of polymer-calcareous coatings

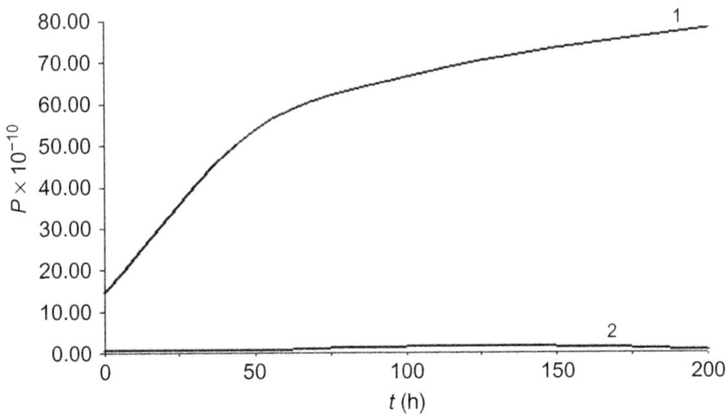

Figure 5.15 Change in the probability of cracking during thermal aging:
1—PVAC coating
2—polymer-calcareous coating

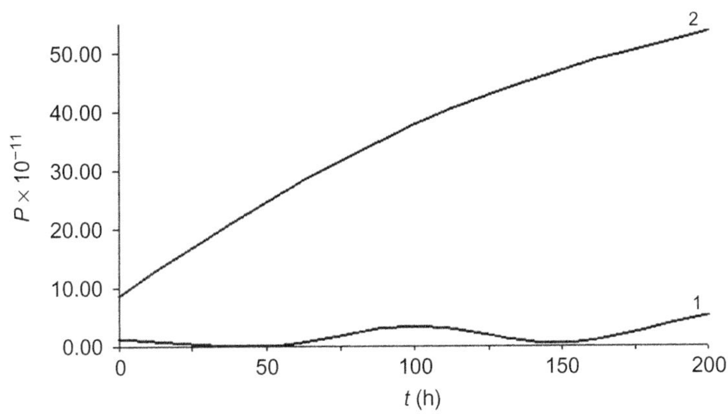

Figure 5.16 Change in the probability of cracking during UV irradiation:
1—PVAC coverage
2—polymer-calcareous coating

References

[1] E. Shindovsky, O. Shyurts, Statisticheskie management techniques kachestvom, Peace, Moscow, 1976.
[2] I.R. Burr, Statistical Quality Control Methods, Marcel Dekker, New York, 1976.
[3] H.G. Charbonneau, G.L. Webster, Industrial Quality Control, Prentice Hall, Englewood Cliffs, NJ, 1978.
[4] GOST R ISO 21747-2010 Statistical methods, Process Performance and Capability Statistics for Measured Quality Characteristics, Standartinform, Moscow, 2012.

[5] V.I. Loganina, The influence of surface quality of coatings on their deformation properties, Contemp. Eng. Sci. 7 (36) (2014) 1935–1941. Available from: https://doi.org/10.12988/ces.2014.411241. HIKARI Ltd, www.m-hikari.com.

[6] V.I. Loganina, Economic estimation of quality process of coloring building products and designs, Contemp. Eng. Sci. 8 (2) (2015) 71–75. Available from: https://doi.org/10.12988/ces.2015.412248. HIKARI Ltd, www.m-hikari.com.

[7] V.I. Loganina, The estimation of reliability of protective-decorative coverings, Contemp. Eng. Sci. 8 (2) (2015) 91–95. Available from: https://doi.org/10.12988/ces.2015.412258. HIKARI Ltd, www.m-hikari.com.

[8] V.I. Loganina, Maintenance of quality of paint and varnish coverings of building products and designs, Contemp. Eng. Sci. 7 (36) (2014). Available from: https://doi.org/10.12988/ces.2014.411243. 943-1947HIKARI Ltd, www.m-hikari.com.

[9] GOST R ISO 21747-2010 Statistical methods, Process Performance and Capability Statistics for Measured Quality Characteristics, Standartinform, Moscow, 2012.

[10] Ford Motor Company, Continuing Process Control and Process Capability Ipmrovement, Ford Motor Company, Dearborn, MI, 1984.

[11] G.J. Hahn, Statistical intervals for a normal population Part 1. Tables. Examples and applications, J. Qual. Technol. 2 (1970) 115–125.

[12] A.A. Bogatyrev, Y.D. Filippov, Standardization of Statistical Methods of Quality Management, Publishing Standards, Moscow, 1989. 120 pp.

[13] GOST R 50779.50–95 Statistical methods, Acceptance Quality Control by Variables. General Requirements, Publishing House of Standards, Moscow, 1995.

[14] T.A. Dubrova, The Statistical Forecasting Methods, Publishing Housc of Unity, Moscow, 2003, p. 205.

[15] L.P. Sullivan, Letters, Qual. Progress 18 (1985) 7–8.

[16] P. T. Jessup, Process Capability, The Value of Improved Performance, in: Paper Presented at the ASQC Automotive Division Workshop Seminar, Southfield, MI, 2–4 November 1983.

[17] J.M. Juran, F.M. Gryna, Quality Planning and Analysis, McGraw-Hill, New York, 1980.

[18] L.P. Sullivan, Reducing variability: a new approach to quality, Qual. Progress 17 (1984) 16–21.

Development of plans for statistical acceptance of quality control

6

6.1 Criteria for acceptance of painted surfaces

For facades of buildings found wide use colorful compositions. Growing competition in the markets of finishing materials along with increasing demands of consumers require manufacturers to provide high-quality painted surfaces. However, as has been seen, finishes are often low quality finish and lead to unplanned repairs and additional costs. Current regulatory and technical literature is devoted mainly to coatings on metal substrates. Study of the regularities of formation of structure and properties of coatings on porous substrates and development of recommendations to improve quality will increase the quality of protective and decorative coatings and ensure maintenance-free lifetimes. Coatings used for the facades of buildings, performing aesthetic protective functions, should have a high-quality appearance, that is, it should be free of defects (inclusions, stains, sharkskin, strokes and scratches, waviness).

In accordance with the statistical theory of strength of solids, the probability of failure of coatings is determined by the presence and concentration of defects, including on surface coatings [1−3]. Thus the quality of coatings, among other factors, determines the resistance of coatings to fracture. Analysis of scientific-technical and normative literature suggests that there is little to no information about the acceptance of quality control rules regarding the painted surfaces of cement concrete. In this regard, development of a methodology for the quality control of the painted surfaces of building products and structures and methods of control is an important scientific-technical and economic problem. The solution to this problem in general will help to increase the service life of protective and decorative coatings.

The quality of any painted surface can be characterized by the class score, a quantitative indicator or any other method. All these methods determine quality by the number and size of defects in the surface area. Each defect characterizes a particular property of the coating, which is the subject of study and control.

Summarizing all the previously mentioned (and reviewed) methods, we can distinguish the following types of defects that determine the set of properties (x_1,x_2,...,x_n): color change (x_1); gloss change (x_2); chalking (x_3); hold dirt (x_4); waviness (x_5); inclusion (x_6); streaks (x_7); strokes, risks (x_8); color variation (x_9); weathering (x_{10}); crack (x_{11}); peeling(x_{12}); dissolution (x_{13}); wrinkling (x_{14}); and

Increasing the Durability of Paint and Varnish Coatings in Building Products and Construction.
DOI: https://doi.org/10.1016/B978-0-12-817046-5.00006-1

bubble (x_{15}) [4–6]. Estimating each of the 15 properties, and summarizing the results, it is possible to obtain comprehensive information about the quality of the coating. The surface will be considered defective if within the area the numerical value of the integral indicator of the quality Q_{coat} is below the specified value Q_{estab}, that is,

$$Q_{\text{coat}} < Q_{\text{estab}} \tag{6.1}$$

$$Q_{\text{coat}} = \sum_{i=1}^{15} a_i \times P_{x_i^{\text{coat}}} \tag{6.2}$$

$$Q_{\text{estab}} = \sum_{i=1}^{15} a_i \times P_{x_i^{\text{estab}}}$$

where a_i is the weighting factors of the i property and $P_{x_i^{\text{coat}}}$ and $P_{x_i^{\text{estab}}}$ are the assessment of real and established indicators of quality properties of the coating relative to the selected base reference defined in general form as

$$P_{x_i^{\text{coat}}} = \frac{x_i^{\text{coat}}}{x_i^{\text{bas}}},\ P_{x_i^{\text{estab}}} = \frac{x_i^{\text{estab}}}{x_i^{\text{bas}}}, \tag{6.3}$$

where x_i^{coat}, x_i^{estab}, x_i^{bas} are the real set and the basic indicators of the quality of coating expressed in any quantitative form.

The solution to the problem of establishing weighting factors for the properties of the coatings were studied by the method of expert assessment. A measure of the consistency of experts adopted the coefficient of concordance W.

It is possible to categorize structures as follows:

I—temporary structures (garages, sheds, outbuildings construction sites, etc.)

II—structures of industrial and civil construction (residential houses, industrial buildings, institutions, municipal administration, etc.)

III—monuments, historical monuments, theaters, government buildings, etc.

The value of the integral indicator of the quality Q_{estab} calculated by the formulae (6.2) and (6.3) are given in Table 6.1.

Computing a quantitative value of the indicator Q_{coat} and comparing the obtained values with Q_{estab}, a conclusion can be made about the quality of the

Table 6.1 Indicators of the quality of the painted surface depending on the type of building

The status of buildings		**I**	**II**	**III**
Values Q_{estab}	Upon acceptance	0.8742	0.9678	1.0
	Critical in the operation	0.309	0.532	0.703

painted surface. To a greater extent the developed technique extends on the surfaces of the external facades of buildings when carrying out an acceptance inspection on the factory, as well as planning for repair works in the process of operation.

6.2 Statistical control of painted surfaces

6.2.1 *Plan of statistical acceptance control of the painted surfaces with constant standard deviation*

The standard deviation s is considered constant and is determined from the condition that "the area in a satisfactory condition coverage" ($Q_{coat} = Q_{estab} \dots 1$) contains six sigmas of the distribution (Fig. 6.1) [7–11]. Thus

$$\sigma = \frac{1 - Q_{estab}}{6} \tag{6.4}$$

Statistical acceptance control of painted surfaces by the quantitative trait is determined by the sample size n (in this case, "sample unit" refers to a defined area of the surface under inspection) and the regulatory level of defects NQL, which is a criterion for inspection. By getting by according control samples of evaluation for the average values of the indicator Q_{coat} by the formula:

$$\overline{Q}_{coat} = \frac{\sum_{i=1}^{n} Q_{coat_i}}{n} \tag{6.5}$$

where n is the number of the controlled areas and comparing them with NQL, a decision is made about the compliance or noncompliance of the painted surface.

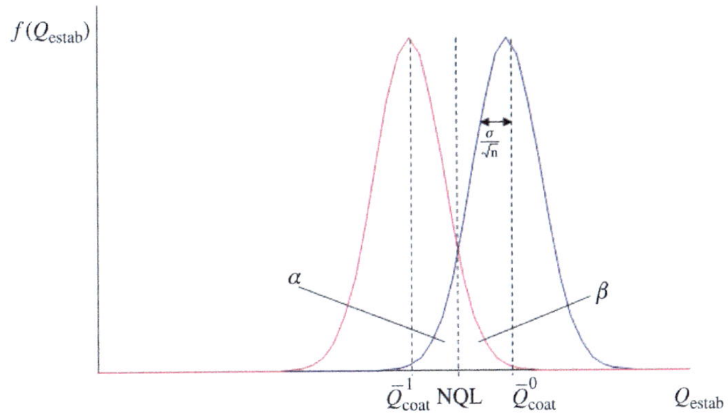

Figure 6.1 The laws of distribution of mean values Q_{coat} of "good" and "bad" coatings.

The distribution of the average Q_{coat} for "bad" and "good" coating(s) will have the form shown in Fig. 6.1.

The values of α and β characterize, respectively, the risks of "vendor" and "consumer" of the painted surfaces.

The values of α and β characterize, respectively, the risks of the "supplier" and "consumer" of painted surfaces. Starting from Fig. 6.1, we can construct a system of equations

$$\begin{cases} \text{NQL} = \overline{Q}^0_{coat} - u_{1-\alpha} \times \dfrac{\sigma}{\sqrt{n}} \\ \text{NQL} = \overline{Q}^1_{coat} + u_{1-\beta} \times \dfrac{\sigma}{\sqrt{n}} \end{cases} \tag{6.6}$$

where $u_{1-\alpha}$ and $u_{1-\beta}$ are the quantiles of the standard normal distribution of the levels $(1-\alpha)$ and $(1-\beta)$, respectively.

Solving the system of Eq. (6.6), we determine the sample volume n:

$$n = \left(\frac{u_{1-\alpha} + u_{1-\beta}}{\overline{Q}^0_{coat} - \overline{Q}^1_{coat}} \right)^2 \times \sigma^2 \tag{6.7}$$

At the same time, according to Fig. 6.1, we have the system of equations

$$\begin{cases} \overline{Q}^0_{coat} = Q_{estab} + u_{1-p_0} \times \sigma \\ \overline{Q}^1_{coat} = Q_{estab} + u_{1-p_1} \times \sigma \end{cases} \tag{6.8}$$

Solving system (6.8), we find

$$\overline{Q}^0_{coat} - \overline{Q}^1_{coat} = (u_{1-p_0} - u_{1-p_1}) \times \sigma \tag{6.9}$$

Substituting Eq. (6.9) into Eq. (6.7), we eventually obtain

$$n = \left(\frac{u_{1-\alpha} + u_{1-\beta}}{u_{1-p_0} - u_{1-p_1}} \right)^2 \tag{6.10}$$

Thus determine the levels of defects for good and bad coatings p_0 and p_1, as well as the risks α and β, can get plan of statistical acceptance control quality of the paintwork including the sample size (number of controlled sections) and the criterion for the average quality index, NQL (Table 6.2). Taking into account the quantitative estimates of Q_{estab} for different periods of operation (Table 6.1), the formulas for the calculations of $\overline{Q}^0_{\text{coat}}$ and $\overline{Q}^1_{\text{coat}}$, and, respectively, NQL will have the forms shown in Table 6.3.

Table 6.2 Sample volumes (number of controlled sections) depending on the typical values of risk β and α, as well as the levels of defects p_0 and p_1

		Values β at $\alpha=0.01$				**Values β at $\alpha=0.05$**				**Values β at $\alpha=0.1$**				**Values β at $\alpha=0.25$**			
		0.01	**0.05**	**0.10**	**0.25**	**0.01**	**0.05**	**0.10**	**0.25**	**0.01**	**0.05**	**0.10**	**0.25**	**0.01**	**0.05**	**0.10**	**0.25**
Values p_1, %	0.5	492	360	298	205	360	247	196	123	298	196	151	88	205	123	88	41
at $p_0=0.27\%$	1	108	79	65	45	79	54	43	27	65	43	33	19	45	27	19	9
	2	42	31	26	18	31	22	17	11	26	17	13	8	18	11	8	4
	3	27	20	17	12	20	14	11	7	17	11	9	5	12	7	5	3
3 Values p_1, %	2	298	218	180	124	218	150	119	74	180	119	92	53	124	74	53	25
at $p_0=1\%$	3	108	79	65	45	79	54	43	27	65	43	33	19	45	27	19	9
	4	65	48	39	27	48	33	26	16	39	26	20	12	27	16	12	6
	5	47	35	29	20	35	24	19	12	29	19	15	9	20	12	9	4
3 Values p_1, %	3	671	489	405	278	489	337	267	167	405	267	206	119	278	167	119	56
at $p_0=2\%$	4	226	165	137	94	165	114	90	57	137	90	70	40	94	57	40	19
	6	87	64	53	36	64	44	35	22	53	35	27	16	36	22	16	8
	8	52	38	32	22	38	26	21	13	32	21	16	10	22	13	10	5
Values p_1, %	4	1285	938	776	533	938	645	512	319	776	512	394	228	533	319	228	107
at $p_0=3\%$	6	213	155	128	88	155	107	85	53	128	85	66	38	88	53	38	18
	9	75	55	45	31	55	38	30	19	45	30	23	14	31	19	14	7
	12	45	33	27	19	33	23	18	11	27	18	14	8	19	11	8	4

Table 6.3 Formulas for calculations $\overline{Q}^{0}_{coat}$, $\overline{Q}^{1}_{coat}$ and NQL for quality control of coatings

The status of buildings		Formulas for calculations $\overline{Q}^{0}_{\text{coat}}$, $\overline{Q}^{1}_{\text{coat}}$ and NQL at the quality control of coatings				
Reception	For all constructions	$Q_{\text{estab}} = 0.971$	$\sigma = 0.0047$	$\overline{Q}^{0}_{\text{coat}} = 0.9716 + u_{1-p_0} \times 0.0047$	$\overline{Q}^{1}_{\text{coat}} = 0.9716 + u_{1-p_1} \times 0.0047$	$\text{NQL} = \overline{Q}^{0}_{\text{coat}} - u_{1-\alpha} \times \frac{0.0047}{\sqrt{n}}$ $\text{NQL} = \overline{Q}^{1}_{\text{coat}} + u_{1-\beta} \times \frac{0.0047}{\sqrt{n}}$
Control during operation	I	$Q_{\text{estab}} = 0.309$	$\sigma = 0.115$	$\overline{Q}^{0}_{\text{coat}} = 0.309 + u_{1-p_0} \times 0.115$	$\overline{Q}_{\text{coat}} = 0.309 + u_{1-p_1} \times 0.115$	$\text{NQL} = \overline{Q}^{0}_{\text{coat}} - u_{1-\alpha} \times \frac{0.115}{\sqrt{n}}$ $\text{NQL} = \overline{Q}^{1}_{\text{coat}} + u_{1-\beta} \times \frac{0.115}{\sqrt{n}}$
	II	$Q_{\text{estab}} = 0.532$	$\sigma = 0.078$	$\overline{Q}^{0}_{\text{coat}} = 0.532 + u_{1-p_0} \times 0.078$	$\overline{Q}^{1}_{\text{coat}} = 0.532 + u_{1-p_1} \times 0.078$	$\text{NQL} = \overline{Q}^{0}_{\text{coat}} - u_{1-\alpha} \times \frac{0.078}{\sqrt{n}}$ $\text{NQL} = \overline{Q}^{1}_{\text{coat}} + u_{1-\beta} \times \frac{0.078}{\sqrt{n}}$
	III	$Q_{\text{estab}} = 0.703$	$\sigma = 0.05$	$\overline{Q}^{0}_{\text{coat}} = 0.703 + u_{1-p_0} \times 0.05$	$\overline{Q}^{1}_{\text{coat}} = 0.703 + u_{1-p_1} \times 0.05$	$\text{NQL} = \overline{Q}^{0}_{\text{coat}} - u_{1-\alpha} \times \frac{0.05}{\sqrt{n}}$ $\text{NQL} = \overline{Q}^{1}_{\text{coat}} + u_{1-\beta} \times \frac{0.05}{\sqrt{n}}$

6.2.2 Statistical acceptance control of quantitative trait with variable standard deviation

In this case, if the standard deviation of the indicators Q_{estab} cannot be considered constant, that is, it changes from sample to sample, for each sample we estimate s using the formula:

$$s = \sqrt{\frac{\sum_{i=1}^{n} \left(Q_{\text{coat}} - \overline{Q}_{\text{coat}}\right)^2}{n-1}} \tag{6.11}$$

In this case, in accordance with Fig. 6.1, the coating is considered valid if the following inequality is observed [12–14]:

$$Q_{\text{estab}} \le \overline{Q}_{\text{coat}} - u_{1-p_1} \times s \tag{6.12}$$

where u_{1-p_1} is the quantile of the standard normal distribution $1-p_1$; p_1 is the permissible (critical) level of defects; and s is the estimate of standard deviation of the studied sample.

Thus from Eq. (6.8), we have the inequality

$$s \le \frac{\overline{Q}_{\text{coat}} - Q_{\text{estab}}}{u_{1-p_1}} \tag{6.13}$$

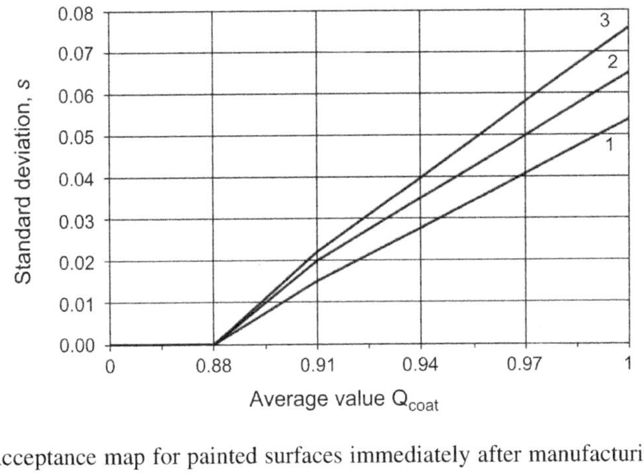

Figure 6.2 Acceptance map for painted surfaces immediately after manufacturing for "status" facilities I:
1—$p_1 = 1\%$
2—$p_1 = 3\%$
3—$p_1 = 5\%$

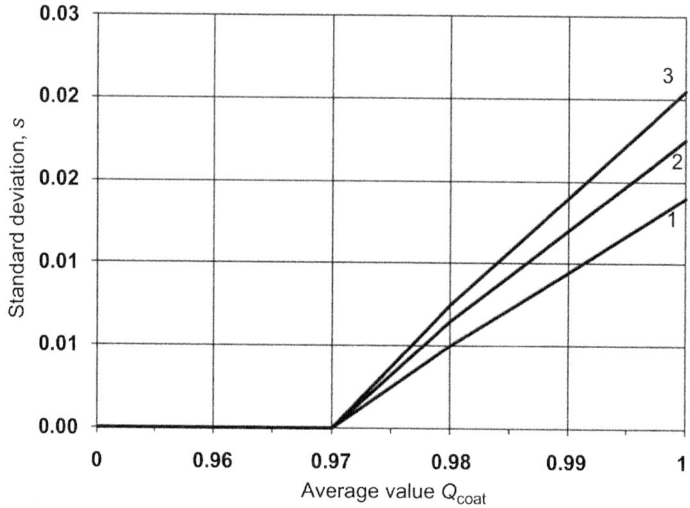

Figure 6.3 Acceptance control map for painted surfaces directly after production for "status" facilities II:
1—$p_1 = 1\%$
2—$p_1 = 3\%$
3—$p_1 = 5\%$

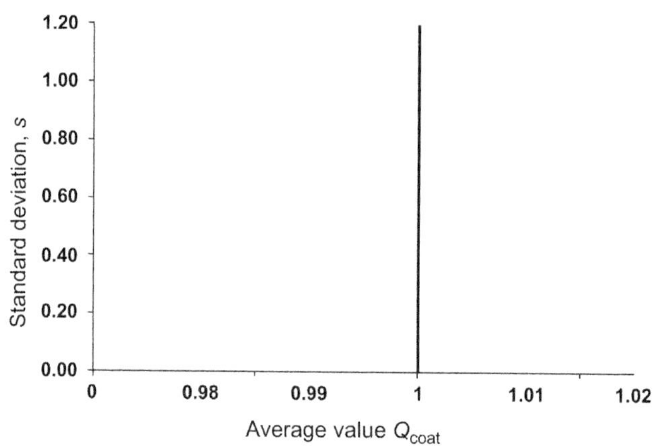

Figure 6.4 Acceptance map for painted surfaces immediately after manufacturing for "status" III facilities for all specified levels of noncompliance.

For each sample the mean value estimate of standard deviation s should be determined. The point with coordinates (Q_{coat}, s) is applied to the acceptance test map and if the point is below the control range, the coating is corresponding, if above, noncorresponding (Figs. 6.2–6.7) [15–17].

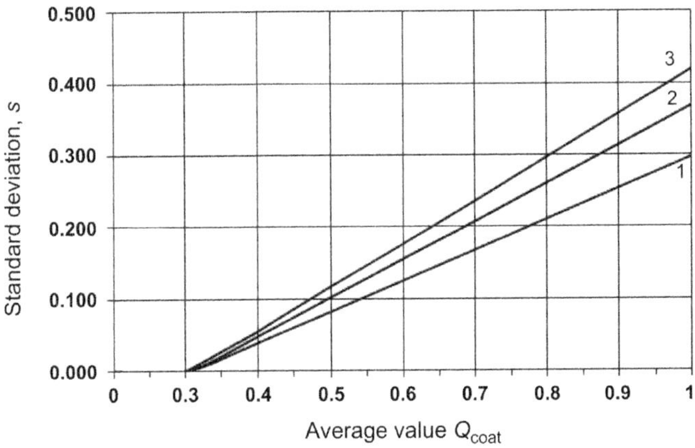

Figure 6.5 Control map for painted surfaces construction "status" I in the process of operation:
1—$p_1 = 1\%$
2—$p_1 = 3\%$
3—$p_1 = 5\%$

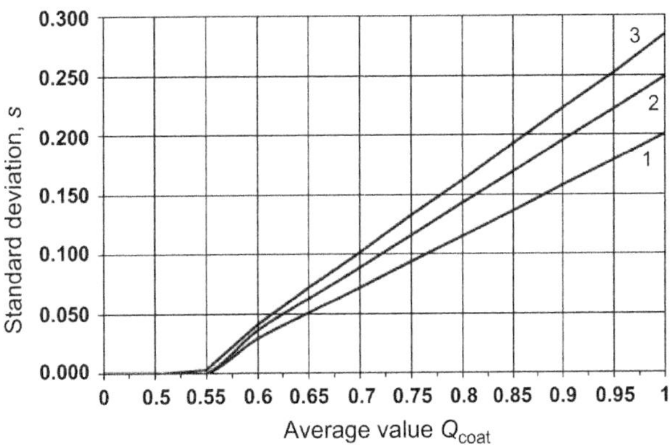

Figure 6.6 Control map for painted surfaces construction "status" II in the process of operation:
1—$p_1 = 1\%$
2—$p_1 = 3\%$
3—$p_1 = 5\%$

6.2.3 Control of the quantification of individual properties

The resistance and actual life of protective and decorative coatings often do not correspond to those forecasted. One of the reasons for this discrepancy is lack of proper control over the painted surface quality, especially concrete and plaster ones, which have a higher number of surface defects compared to metal ones [18–20].

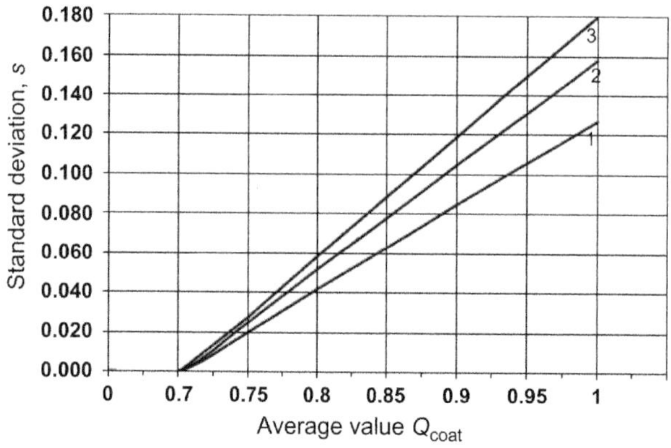

Figure 6.7 Control map for painted surfaces facilities "status" III during operation:
1—$p_1 = 1\%$
2—$p_1 = 3\%$
3—$p_1 = 5\%$

As is known, the variability of paint and varnish materials properties follows the normal law of distribution. Assuming, that the levels of the discrepancies of the coating protective and decorative properties parameters make q_1 and q_2. The probability that the painted surface will be good according to both parameters is equal:

$$P = (1 - q_1)(1 - q_2) \tag{6.14}$$

Eq. (6.14) corresponds to a production without defects.

A method of statistical acceptance control of painted surfaces construction and products is proposed. The technique is based on the control of the particular areas of the surface. The number of areas is determined by calculating.

This technique is based on the definition of average and standard deviation (SD) of quantitative assessments of different quality parameters and on the calculation of the real defect level (percentage of poor areas of the surface) according to each parameter.

The quality of the painted surface is evaluated with the quantitative assessments of decorative and protective properties:

- shine change
- color change
- mud retention
- chalking
- cracking
- scaling
- weathering
- bubbles formation

The value of the generalized assessment of decorative of the properties coatings is calculated by the formula

$$AD = Xa\text{Ц} + Xa\text{В} + Xa\text{М} + Xa\Gamma \tag{6.15}$$

where X is the weighting factor of each property.

The value of the generalized estimation of protective properties of coatings $A3$ is calculated by the formula

$$A3 = X(0.6aT + 0.4a\text{ЛР}) + X(0.6aB + 0.4a\text{ЛР}) + X(0.6a\Pi + 0.4a\text{ЛР}) + X(0.6aC + 0.4a\text{ЛР}) \tag{6.16}$$

where X is the weighting factor of each type of fracture and $a\text{ЛР}$ is the relative estimation of damages (diameter, depth); T is the cracking; B is the weathering; C is the peeling; and Π is the bubbles formation.

There is a set quantitative assessment scale for each parameter depending on the coating condition.

The top border of a good condition of decorative properties of coverings is accepted under condition of $AD = 1$, and the bottom border at $AD = 0.7$.

The top border of a good condition of protective properties of coverings is accepted under condition of $A3 = 1.0$, and the bottom border at $A3 = 0.76$.

Consequently, the main requirement, which will determine other requirements, is the requirement for the quality of the painted surface as a whole, formulated as follows: "The percentage of poor surface should not exceed $q\%$."

The solution to the problem of determining the defect levels for a particular area is as follows.

Assuming the quality of the painted surface is characterized by m properties, the probability that the surface will be good according to all the parameters is defined as follows:

$$P = (1 - q) = (1 - q_1) \times (1 - q_2) \times \ldots \times (1 - q_m) \tag{6.17}$$

where $q_1, q_2, \ldots, q_n$ are the areas of the surface which is poor according to a particular property; and q is the area of the surface that is poor according to all the properties.

Expression (6.13) corresponding to the proportion of all surface quality parameters in control, obviously, is transformed into the inequation:

$$P = (1 - q) > (1 - q_1) \times (1 - q_2) \times \ldots \times (1 - q_m) \tag{6.18}$$

Eq. (6.18) provides the criteria to accept or reject the painted surface.

Let us consider a particular case when all the properties of the coating are equal, that is, $q_1 = q_2 = \ldots = q_m = q^*$. Then, solving inequation (6.18), we can determine the critical levels of discrepancies for each property:

$$q* < 1 - \sqrt[m]{1 - q} \tag{6.19}$$

The possible inequation solutions (6.19) are shown in Table 6.4.

Being guided by the manufacturer and consumer risks α and β (tolerable alpha and beta errors), and also critical levels of discrepancies for good and poor coatings (q_0 and q), we determine the sample number (number of controlled areas of a surface) by the formula [7,11]

$$n = \left(\frac{u_{1-\alpha} + u_{1-\beta}}{u_{1-q_0} - u_{1-q_1}} \right)^2 \tag{6.20}$$

where $u_{1-\alpha}$, $u_{1-\beta}$, u_{1-q_0}, u_{1-q_1}, are the quantiles of the standard normal distribution of corresponding levels.

Having drawn random samples from n areas of the painted surface we determine the quantitative assessment of specified properties for each area:

$$\overline{S_i} = \frac{\sum_{j=1}^{n} S_i^j}{n} \tag{6.21}$$

$$\sigma_{S_i} = \sqrt{\frac{\sum_{j=1}^{n} \left(S_i^j - \overline{S_i}\right)^2}{n-1}} \tag{6.22}$$

where S_i^j is a quantitative assessment of property i on area j and n is number of areas.

The assessment and calculation results are shown in Table 6.5.

Then the real defect level for each property is calculated by

$$q_i = 1 - \Phi\left(\frac{\overline{S_i} - S_{\kappa p_i}}{\sigma_{S_i}} \right) \tag{6.23}$$

where $S_{\kappa pi}$ is the set critical value of i property of the coating and $\Phi(x)$ is the value of the normal standard distribution function.

Table 6.4 Critical levels of the coating discrepancies for a particular property (q^*)

Number of quality parameters m	Determined share of defective surface		
	0.01	**0.05**	**0.1**
2	5.013×10^{-3}	0.025	0.051
4	2.509×10^{-3}	0.013	0.026
6	1.674×10^{-3}	8.512×10^{-3}	0.017
8	1.256×10^{-3}	6.391×10^{-3}	0.013
10	1.005×10^{-3}	5.116×10^{-3}	0.01

Table 6.5 Quantitative assessments of specific properties of a coating

No. area	No. property				
	1	**2**	**3**	...	**m**
1	S_1^1	S_2^1	S_3^1	...	S_m^1
2	S_1^2	S_2^2	S_3^2	...	S_m^2
3	S_1^3	S_2^3	S_3^3	...	S_m^3
...	...	...	...	...	...
n	S_1^n	S_2^n	S_3^n	...	S_m^n
	$\overline{S_1}$	$\overline{S_2}$	$\overline{S_3}$	...	$\overline{S_m}$
	σ_{S_1}	σ_{S_2}	σ_{S_3}	...	σ_{S_m}

Having defined the real values q_i for properties, we compare them with the values specified by the requirements and draw conclusions about the quality of a coating by particular properties.

If the requirements specified the quality of the coating as a whole (by all the properties), then we determine the value q and either accept or reject the coating.

6.3 Reliability and prediction of properties of protective and decorative coatings

6.3.1 Criteria of reliability of paint and varnish coatings

Forecasting the service life of protective and decorative coatings is of great practical importance, since it allows for effective planning of current and future repairs. The development of mathematical models characterizing the processes of aging (wear) of coatings allows, on the basis of a comparative analysis, to carry out scientifically based selection of the paint composition and coating application technology (in accordance with operating conditions and customer requirements). No doubt, effective solutions to the above tasks are decisive in ensuring the quality of coatings.

With this in mind, it is necessary to establish the determining criteria for these properties and the concretization of the quantitative concept of "failure" from the standpoint of reliability theory.

The "failure" of a coating can be characterized in two ways:

- Decrease in the integral quality indicator Q_{coat} below the value Q_{estab} set for a given period of time
- Reduction of quantitative estimates of individual coating properties (the most critical) is lower than the allowable for a given period of time Q_{estab}^i

The "refusal" of coverage in the second case will be interpreted in a similar way, only instead of the integral indicator Q_{estab} a particular indicator should be used Q_{estab}^i.

The reasons for the formation of defects on the surface of the coatings can be divided into three groups: technological, constructive and operational.

Technological reasons for the formation of defects is as follows. When finishing the facades of buildings under construction conditions, various deviations from the requirements of the painting technology are often possible, which leads to the appearance of defects. Thus the widespread defect of furnish is a delamination of a covering together with plaster. In addition, due to the poor preparation of the substrate (e.g., the application of an equalizing layer of various thicknesses), in the first year of operation, the coating heterogeneity is observed.

The appearance of contamination of coatings in the form of wet spots is mainly due to imperfection of the constructions of window and visors of entrances. In these places, as early as the second year of operation, cracks and delamination of the coatings appear.

In the process of operation, the resistance of coatings is significantly influenced by temperature deformations caused by different linear coefficients of thermal linear expansion of coatings and concrete.

The criteria for the reliability of paintwork should be:

- Probability of failure-free operation during a specified period of operation (the established interval between planned repairs)
- Probability of failure
- Failure rate
- Mean time of trouble-free operation

The probability of failure-free operation of the paintwork will be the probability that under the given operating conditions and the specified time interval there will be no decrease in the integral (or private) quality index of the coating below the established level.

We denote this characteristic by $P(Q_{\text{coat}} < Q_{\text{estab}}|t)$.

According to definition

$$P\left(Q_{\text{coat}} < Q_{\text{estab}}\middle|t\right) = p(T_1 > t) \tag{6.24}$$

where t is the time during which the probability of failure-free operation of the coating is determined and T_1 is the time of the coverage from the moment of its creation to the recognition of the coverage of the denied.

The probability of failure-free operation in statistical studies of the reliability of coatings should be evaluated by the expression:

$$P\left(Q_{\text{coat}} < Q_{\text{estab}}\middle|t\right) = \frac{N_0 - n(t)}{N_0} \tag{6.25}$$

where N_0 is the number of coatings to be examined (sections of the same coating) and $n\ (t)$ is the number of the refused products in time t.

$P(Q_{coat} < Q_{estab}|t)$ is a statistical estimate of the probability of failure-free operation of coatings.

At a large number N_0, the statistical estimate $P(Q_{coat} < Q_{estab}|t)$ practically coincides with the probability of failure-free operation $P(Q_{coat} < Q_{estab}|t)$.

In practice, the probability of failure, which we denote by $Q\ (t)$, is often a more convenient characteristic of the reliability of coatings.

The probability of failure will be the probability that under given operating conditions, at least one failure of the coating sample will occur in a given time interval. The probability of failure and the probability of failure-free operation are inconsistent and opposite events, *i*.

$$Q(t) = 1 - P(Q_{coat} < Q_{estab}|t) \tag{6.26}$$

On the basis of Eqs. (6.24) and (6.26)

$$Q(t) = p(T_1 \leq t) \tag{6.27}$$

It follows from Eq. (6.27) that the failure probability is an integral function of the time distribution of the coverage T_1 to the first failure, that is,

$$Q(t) = F(t) \tag{6.28}$$

The derivative of the integral distribution function is the differential law (density) of the distribution $f\ (t)$. On the basis of expressions (6.25) and (6.26), the statistical estimate of the probability of failure is the expression:

$$Q * (t) = \frac{n(t)}{N_0} \tag{6.29}$$

The frequency of failure of the coating will be the ratio of the number of refused coatings (coating areas) per unit time to the initial number of coatings under investigation (coating sites), provided that all failed coatings (coating areas) are not restored. We denote this characteristic by $a\ (t)$.

According to the definition

$$a(t) = \frac{n(t)}{N_0 \times \Delta t} \tag{6.30}$$

where $n\ (t)$ is the number of refused coating samples in the time interval from $t - \frac{\Delta t}{2}$ to before $t + \frac{\Delta t}{2}$.

The probabilistic definition of this characteristic is derived through the definition of the analytic relationship between $a\ (t)$ and $P(Q_{coat} < Q_{estab}|t)$.

The number of refused samples $n\ (t)$ in expression (6.30) is equal to the difference in the number of suitable coatings (coating areas), at the beginning and end of the interval Δt, that is,

$$n(t) = N(t) - N(t + \Delta t) \tag{6.31}$$

where $N\ (t)$ is the number of samples at the beginning of the interval Δt and $N\ (t + \Delta t)$ is the number of samples at the end of the interval Δt.

On the basis of Eq. (6.25):

$N(t) = N_0 \times P(Q_{coat} < Q_{estab}|t)$; $N(t + \Delta t) = N_0 \times P(Q_{coat} < Q_{estab}|t + \Delta t)$.

After substituting the values of $n\ (t)$ in Eq. (6.31), we obtain:

$$a(t) = P(Q_{coat} < Q_{estab}\Big|t) - P(Q_{coat} < Q_{estab}\Big|\frac{t + \Delta t)}{\Delta t} \tag{6.32}$$

Letting Δt go to zero and passing to the limit, we obtain

$$a(t) = -P'(Q_{coat} < Q_{estab}|t) = Q'(t) \tag{6.33}$$

It follows that the failure rate is the probability density (or the distribution law) of the uptime of the coating until the first failure. Expression (6.33) is a probabilistic definition of the failure rate of coatings.

On the basis of Eq. (6.33) we have

$$\begin{aligned} &Q(t) = \int_0^t a(t)dt, \\ &P(Q_{coat} < Q_{estab}|t) = 1 - \int_0^t a(t)dt\} \end{aligned} \tag{6.34}$$

The average time of nonfailure operation of the coverage will be referred to as the mathematical expectation of the "work" time of coating to failure. This characteristic will be denoted t_{aver}. The mathematical expectation t_{aver} can be determined by the failure rate:

$$t_{aver} = \int_{-\infty}^{+\infty} t \times a(t)dt \tag{6.35}$$

Since t is positive,

$$t_{aver} = \int_0^{+\infty} t \times a(t)dt = -\int_0^{+\infty} t \times P'(Q_{coat} < Q_{estab}|t)dt \tag{6.36}$$

Definition of the considered criteria of reliability of coatings is impossible without definition of the law of distribution of time of nonfailure operation or time of failure, that is, analytical dependence $P(Q_{coat} < Q_{estab}|t)$ or $Q(t)$.

The established reliability criteria are universal in the sense that they allow comparative analysis of the properties of the coatings and to calculate the "repair intervals" depending on the requirements set by the consumer, expressed as an integral quality indicator Q_{estab} or in the form of partial indicators for individual properties Q^i_{estab}.

6.3.2 Law of probability distribution of trouble-free service of protective and decorative coatings

The urgency of the task of developing a universal law for the distribution of the probability of failure-free operation of coatings (or the distribution of failure rates) is the need to determine "overhaul" intervals and predict the life of protective and decorative coatings. The application of this law in practice will make it possible to judge objectively on the basis of the accumulated statistical data on the possibility of meeting the requirements of consumers.

Obviously, like any other, the law of distribution of a random variable (and the fact that the time to failure of coverage is a random variable is unquestionable), having an applied character (indicative, Poisson, Weibull, etc.), the law should be determined by the parameters, which have some physical meaning. In this problem, such parameters will be:

- time to failure t_{aver};
- time interval of aging (wear), Δt; and
- the shape of the curve describing the probability of failure in the aging time interval, which is characterized by the failure rate $a\ (t)$.

The well-known and widely used theory of reliability laws of distribution of the operating time to failure (normal, exponential, Weibull) explicitly do not provide these requirements.

Normal distribution law

$$f(x) = \frac{1}{\sqrt{2\pi\sigma}} \times e^{-\frac{(x-\bar{x})^2}{2\sigma^2}} \tag{6.37}$$

determines the position ($\bar{x}$) and scattering (σ) of the value of the exponent, but does not determine the behavior of the variability of the exponent in the scattering interval (in general, the variability of the exponent in the scattering interval using the normal law can be characterized by the indicators of asymmetry and kurtosis; however, in this case, this is not convenient).

Exponential distribution

$$P(t) = 1 - e^{-\lambda t} \tag{6.38}$$

where λ is the failure rate, characterizing the aging rate, taking into account the variability of λ during operation and the approximation error, which does not always allow to recognize the model as adequate.

Weibull distribution

$$P(t) = 1 - e^{-\left(\frac{t-t_0}{\beta}\right)^{\alpha}} \tag{6.39}$$

where t_0, β, α are the parameters of the shift, scale, shape, respectively, and more accurately describe the behavior of the probability of failure of coatings, but its use presents certain difficulties in conducting a comparative analysis, because the parameters α and β do not lend themselves to physical interpretation (they characterize the distribution curve, but not the average operating time, the aging interval or other properties of the coating).

Nevertheless, we take the Weibull distribution as the basis for the law being developed, assuming that the parameter β will characterize the position of the characteristic under study (the mean time of failure), and α is the aging rate. The initial time t_0 is assumed to be zero. In this case, the Weibull distribution, which determines the probability of failure of the coating, is represented as

$$Q(t) = 1 - e^{-\left(\frac{t}{\bar{t}}\right)^s} \tag{6.40}$$

where $\bar{t}$ is the mean time of failure, calculated from the experimental data and s is a coefficient characterizing the change in the rate of aging in the time interval in which failures are observed.

Proceeding from the above assumptions, the coefficient s is determined by equating the first derivative of the function (6.40) at a point $\bar{t}$ to the value $1/\Delta t$, where Δt is the time interval between the first and last failures of the cover quantities under consideration.

We get:

$$Q'(\bar{t}) = \frac{s}{\bar{t} \times e} = \frac{1}{\Delta t} \tag{6.41}$$

whence

$$s = \frac{\bar{t} \times e}{\Delta t} \tag{6.42}$$

and consequently, the distribution law (6.39) takes the form

$$Q(t) = 1 - e^{-\left(\frac{t}{\bar{t}}\right)^{\frac{\bar{t}e}{\Delta t}}} \tag{6.43}$$

Let us consider the behavior of the function (6.43) for fixed values of $\bar{t}$ and Δt.

Fig. 6.8 shows the graphs of the dependence of the probability of the onset of failure of the operating time (the operating time in this case will be characterized as the number of humidification-drying cycles for coating) for a fixed value of Δt ($\Delta t = 200$) and different $\bar{t}$ ($\bar{t} = 100, 300, 500, 700, 900$). Analysis of the curves in Fig. 4.1 allows us to conclude that the value $\bar{t}$ uniquely determines the position of the curve and its change does not affect the form of the dependence.

Fig. 6.9 shows the graphs of the dependence of the probability of the onset of failure of the operating time for a fixed value of $\bar{t}$ ($\bar{t} = 200$) and different Δt ($\Delta t = 100, 300, 500, 700, 900$). Analysis of the curves allows us to conclude that the value uniquely determines the aging time of the coating (the time interval from the probability of failure to $Q(t) \approx 0$ to $Q(t) \approx 1$).

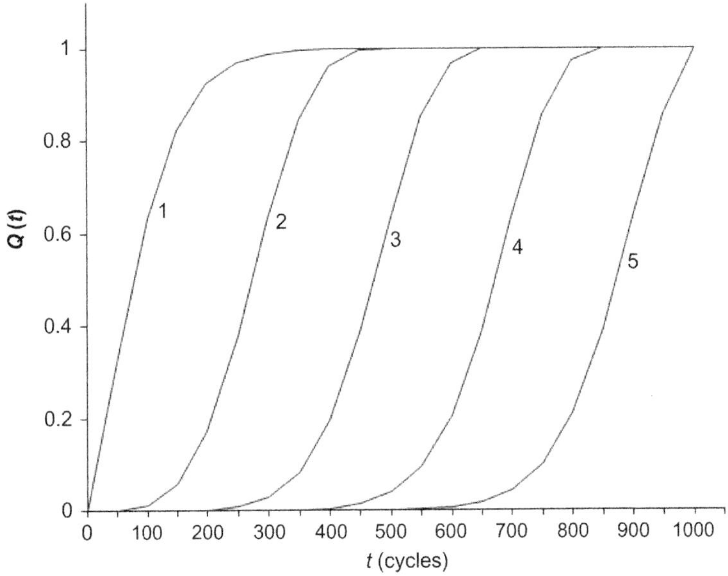

Figure 6.8 The probability of failure of coverage at $\Delta t = 200$:
1—$\bar{t} = 100$ cycles
2—$\bar{t} = 300$ cycles
3—$\bar{t} = 500$ cycles
4—$\bar{t} = 700$ cycles
5—$\bar{t} = 900$ cycles

The curves in Figs. 6.8 and 6.9, the data allow us to conclude that the shape of the distribution curve in the aging time interval of the coating will be determined by the ratio $\bar{t}/\Delta t$.

The probability of failure-free operation of the coating $P(Q_{coat} < Q_{estab}|t)$ according to Eq. (6.26), will be calculated as

$$P(Q_{coat} < Q_{estab}|t) = 1 - Q(t) = 1 - \left(1 - e^{-\left(\frac{t}{\bar{t}}\right)^{\frac{\bar{t}\times e}{\Delta t}}}\right) = e^{-\left(\frac{t}{\bar{t}}\right)^{\frac{\bar{t}\times e}{\Delta t}}} \tag{6.44}$$

The probability density (the distribution law) of the nonfailure operation time of coatings (failure rate a (t)) will be as follows:

$$a(t) = \frac{\bar{t}\times e}{t\times \Delta t} \times \left(\frac{t}{\bar{t}}\right)^{\frac{\bar{t}\times e}{\Delta t}} \times e^{-\left(\frac{t}{\bar{t}}\right)^{\frac{\bar{t}\times e}{\Delta t}}} \tag{6.45}$$

Figs. 6.10 and 6.11 shows the graphical representations of the probability densities of fail-safe time for a fixed value of Δt ($\Delta t = 200$) and various $\bar{t}$ ($\bar{t} = 100$, 300, 500, 700, 900) and for a fixed value of $\bar{t}$ ($\bar{t} = 200$) and various ($\Delta t = 100$, 300, 500, 700, 900).

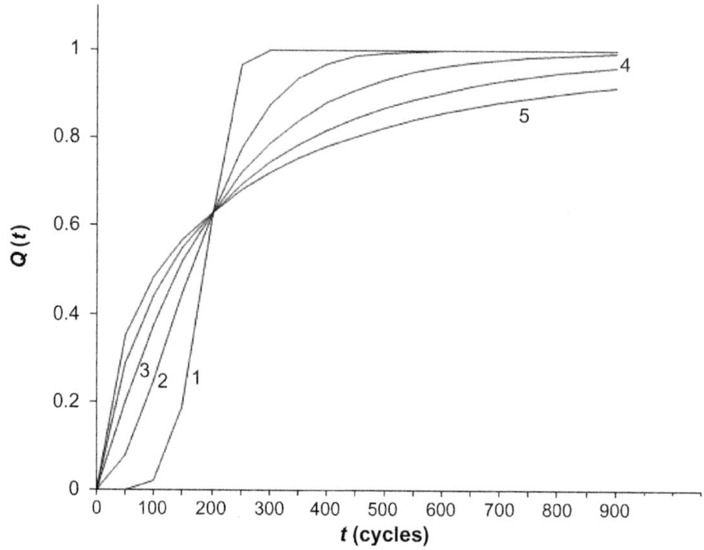

Figure 6.9 The probability of coating failure at $\bar{t} = 200$ cycles:
1—$\Delta t = 100$ cycles
2—$\Delta t = 300$ cycles
3—$\Delta t = 500$ cycles
4—$\Delta t = 700$ cycles
5—$\Delta t = 900$ cycles

The mathematical expectation of the failure time t_{aver} is determined according to the expression (6.40):

$$t_{\text{aver}} = \int_0^\infty t \times a(t)dt = \int_0^\infty t \times \frac{\bar{t} \times e}{t \times \Delta t} \times \left(\frac{t}{\bar{t}}\right)^{\frac{\bar{t} \times e}{\Delta t}} \times e^{-\left(\frac{t}{\bar{t}}\right)^{\frac{\bar{t} \times e}{\Delta t}}} dt$$
$$= \int_0^\infty \frac{\bar{t} \times e}{\Delta t} \times \left(\frac{t}{\bar{t}}\right)^{\frac{\bar{t} \times e}{\Delta t}} \times e^{-\left(\frac{t}{\bar{t}}\right)^{\frac{\bar{t} \times e}{\Delta t}}} dt \quad (6.46)$$

Dispersion, respectively, will be calculated as

$$D(t) = \int_0^\infty (t - t_{\text{aver}})^2 \times \frac{\bar{t} \times e}{t \times \Delta t} \times \left(\frac{t}{\bar{t}}\right)^{\frac{\bar{t} \times e}{\Delta t}} \times e^{-\left(\frac{t}{\bar{t}}\right)^{\frac{\bar{t} \times e}{\Delta t}}} dt \quad (6.47)$$

The standard deviation of the distributions is determined, respectively, as

$$\sigma = \sqrt{D(t)} \quad (6.48)$$

The shape of Figs. 6.10 and 6.11 allow us to note that the distributions have some asymmetry, which can be calculated as

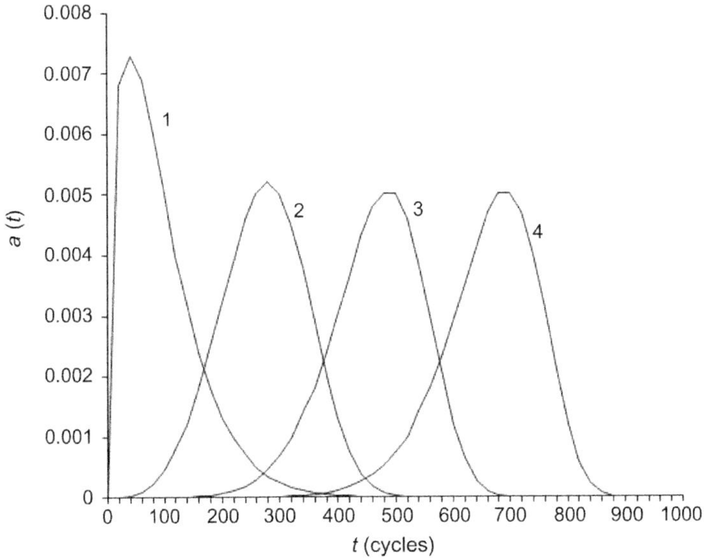

Figure 6.10 Density of probability of failure of a cover at $\Delta t = 200$:
1—$\bar{t} = 100$ cycles
2—$\bar{t} = 300$ cycles
3—$\bar{t} = 500$ cycles
4—$\bar{t} = 700$ cycles

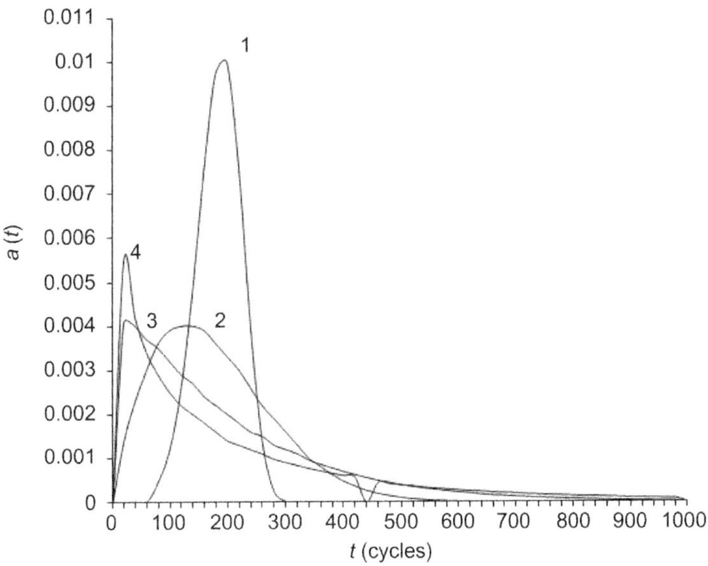

Figure 6.11 The probability density of a coating failure at $\bar{t} = 200$ cycles:
1—$\Delta t = 100$ cycles
2—$\Delta t = 300$ cycles
3—$\Delta t = 500$ cycles
4—$\Delta t = 700$ cycles

$$A = \frac{\mu_3}{\sigma^3} \tag{6.49}$$

where μ_3 is the third central moment of distribution, which is calculated as

$$\mu_3 = \int_0^{\infty} \left(t - t_{cp}\right)^3 \times \frac{\bar{t} \times e}{t \times \Delta t} \times \left(\frac{t}{\bar{t}}\right)^{\frac{\bar{t} \times e}{\Delta t}} \times e^{-\left(\frac{t}{\bar{t}}\right)^{\frac{\bar{t} \times e}{\Delta t}}} dt \tag{6.50}$$

Summarizing the above conclusions, we can conclude that function (6.43) meets all the above requirements for the distribution law, and each of its parameters has a concrete physical interpretation.

In order to confirm the applicability of the developed distribution law, various samples of three types of coatings were tested: PVAC, PVAC with addition of silicone fluid GKZH-94 and polymer lime. The tests consisted of the effect of alternate wetting-drying and observation of the change in the integral quality index of the painted surfaces. Refusal was fixed, if $Q_{coat} < 0.5$.

The results of the tests are given in Table 6.6.

The experimental and theoretical distributions calculated from formulas (6.36) and (6.43) are shown in Figs. 6.12–6.14.

The mathematical expectation of the failure time t_{aver}, calculated from Eq. (6.20) for the coatings under consideration, is given in Table 6.7.

Experimental studies have shown that the application of the distribution law (6.43) makes it possible to describe with high enough accuracy (the probability of the consistency of the experimental distribution with the theoretical criterion χ^2 with the number of degrees of freedom $k = 3$ greater than 0.8) of the probability of the time of failure of the coatings. Under real operating conditions, the time parameter t must be expressed in days, months, years, etc., which will determine the longevity of the coatings in certain climatic conditions.

Thus the results obtained make it possible to recommend the use of the probability distribution function of the operating time to failure as a universal quality tool for estimating the service life of paint and varnish coatings.

Evaluation of the reliability of protective and decorative coatings in the process of operation presents certain difficulties associated with the duration and laboriousness of the tests. This means the problem of recalculating the reliability indicators obtained under forced test conditions into real ones is very actual. Certain difficulties arise due to the fact that, depending on the intensity of the operating factors and the test regime, there is a certain rate of resource consumption (the intensity of the "failure"). The greater the intensity of the external impact and the rigidity of the test regime, the more, over a certain period of time, the system uses a greater resource.

To obtain the conversion functions, we consider two tests:

functioning of the coating in the normal mode of field tests—$I_n(t)$

forced cycle tests—4 hours freezing at a temperature of $-40°$C, 2 hours defrosting in air at a temperature of 40°C and relative humidity of 60%, 2 hours, humidification at a temperature of 18°C–200°C and a relative humidity of 60%–70%—$I_f(t)$

Table 6.6 Changes in the integral quality index of the painted surfaces

Coating	Sample number	Cycles, t									$\bar{t}$	Δt
		50	100	120	130	150	200	250	300	350	230.7	220
PVAC	1						Failure					
	2					Failure						
	3						Failure					
	4				Failure							
	5								Failure			
	6								Failure			
	7					Failure						
	8								Failure			
	9						Failure					
	10						Failure					
	11									Failure		
	12					Failure						
	13								Failure			
	14								Failure			
PVAC with addition of silicone fluid GKZH-94	1									Failure	267.8	200
	2							Failure				
	3					Failure						
	4								Failure			
	5								Failure			
	6						Failure					
	7									Failure		
	8							Failure				
	9								Failure			
	10							Failure				
	11								Failure отказ			
	12						Failure					
	13								Failure			
	14							Failure				

(Continued)

Table 6.6 (Continued)

Coating	**Sample number**	**Cycles, *t***									$\bar{t}$	Δt
Polymer lime	1						Failure					
	2					Failure					192.1	200
	3		Failure									
	4					Failure						
	5						Failure					
	6			Failure								
	7								Failure			
	8						Failure					
	9						Failure					
	10								Failure			
	11			Failure тказ								
	12								Failure			
	13					Failure						
	14						Failure					

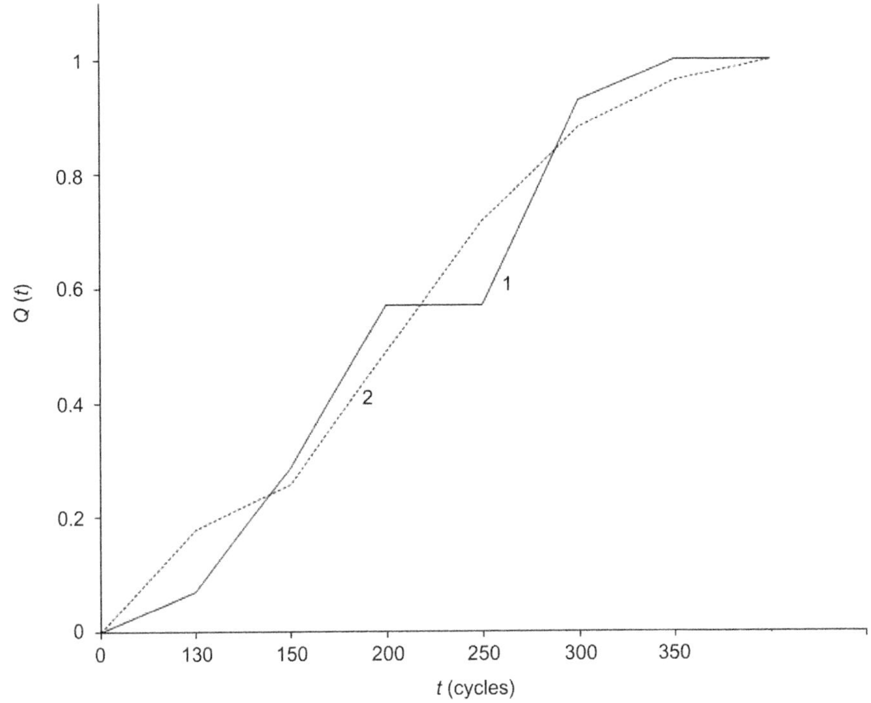

Figure 6.12 The probability of failure of PVAC coverage:
1—experimental distribution
2—theoretical distribution$P(t) = 1 - e^{-\left(\frac{t}{230.7}\right)^{\frac{230.7e}{220}}}$

We denote by $\varepsilon_f(t)$ and $\varepsilon_n(t)$ the rate of change in the value of the determining parameter (e.g., the adhesion strength) in the indicated modes. In accordance with the physical principle of reliability N.M. Sedyakin

$$\int_0^t \lambda_\phi(z)dz = \int_0^{x(t)} \lambda_H(z)dz \tag{6.51}$$

$$\int_0^t \varepsilon_\phi(z)dz = \int_0^{x(t)} \varepsilon_H(z)dz \tag{6.52}$$

where $x(t)$ is the function of time recalculation of fail-safe operation from a mode $I_f(t)$ to the $I_n(t)$ and Δt and $\lambda_f(t)$ is the intensity of failure, respectively, in full-scale and forced tests.

It was shown in Refs. [21,22] that in accelerated tests the change in the determining parameter is accompanied by an acceleration due to the acceleration of the process of degradation of the properties of the system with increasing external influences. The presence of the acceleration of the change in the parameter is a

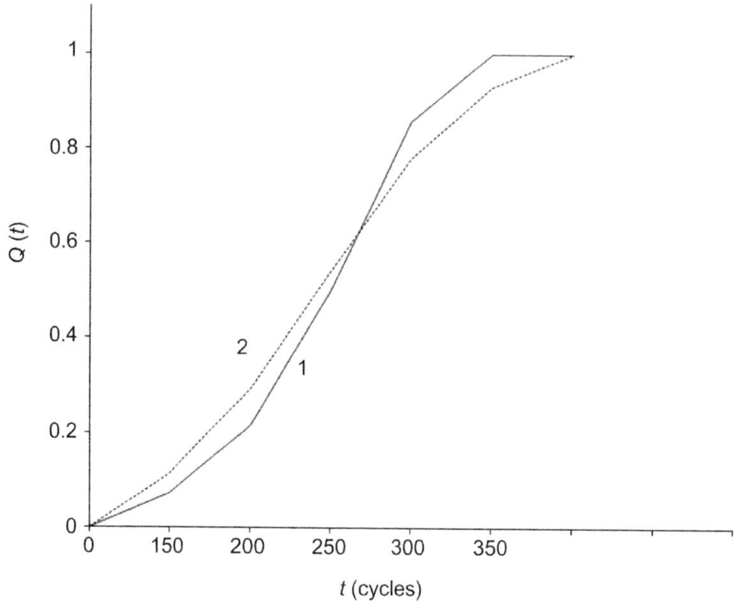

Figure 6.13 The probability of failure of the PVAC coating with the addition of GKZH-94:
1—experimental distribution
2—theoretical distribution$P(t) = 1 - e^{-\left(\frac{t}{267.8}\right)^{\frac{267.8e}{200}}}$

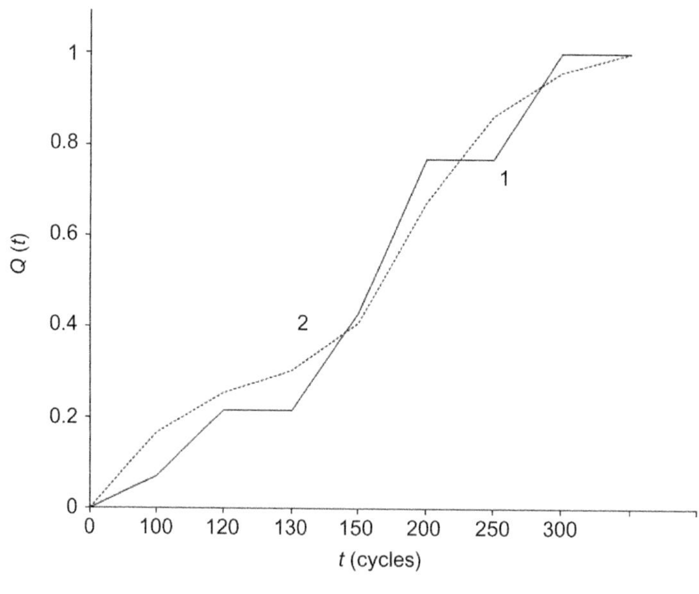

Figure 6.14 The probability of failure of the polymer-lime coating:
1—experimental distribution
2—theoretical distribution$P(t) = 1 - e^{-\left(\frac{t}{192.1}\right)^{\frac{192.1e}{200}}}$

Table 6.7 Mathematical expectation of the failure time

Coating	t_{aver} (cycles)
PVAC	205.57
PVAC with the addition of GKZH-94	241.46
Polymer lime	170.65

necessary condition for the distribution of the time to failure of the article in different modes to be different.

Solving the system of Eqs. (6.51) and (6.52), we find that

$$x'(t) = \frac{\varepsilon_f(t)}{\varepsilon_n(x(t))} \tag{6.53}$$

$$\varepsilon_n(x(t)) = \frac{\varepsilon_f(t)}{x'(t)} \tag{6.54}$$

$$\lambda_n(x(t)) = \frac{1}{x'(t)}\lambda_f(t) \tag{6.55}$$

where $x(t)$ is the conversion function.

Thus it is possible to predict the reliability of coatings when operating in full-scale conditions on the basis of data from forcing tests. For this you need to know the:

Law of change of the determining parameter in the regime $I_f(t)$
Translation function $x(t)$
Law of distribution of time of trouble-free operation in the mode $I_f(t)$

Consider the calculation of reliability using the example of calcareous protective and decorative coatings. As a criterion for the weather resistance of coatings, a change in protective properties was used, evaluated in accordance with GOST 6992-68 on an eight-point system. During the experiment, the protective properties of the coatings were evaluated, as well as the adhesion strength. The total number of tests was 50 operating cycles. The obtained data were compared with the data of field surveys. The results are shown in Fig. 6.15.

The results of the studies show that the change in protective properties in full-scale tests corresponds to an exponential dependence of the form

$$Y(t) = A\exp(-\alpha t) \tag{6.56}$$

Then the rate of change is

$$\varepsilon_f(t) = -\alpha \times A\exp(-\alpha t)$$

For the calcareous coating in question $A = 7.96;\ \ \alpha = 0.02$.

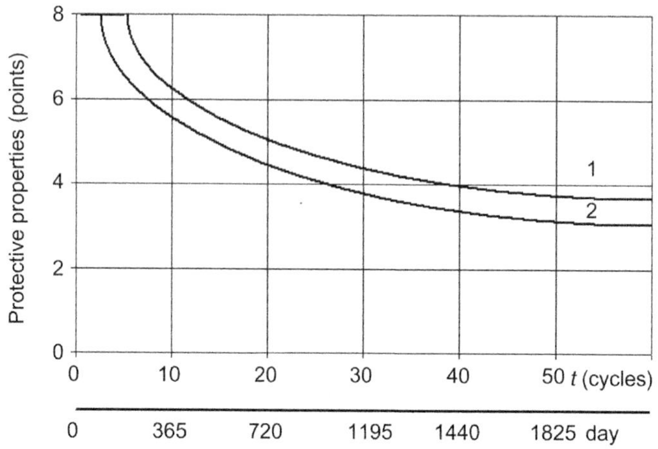

Figure 6.15 Change in the protective properties of the lime coating in the process of aging
1—forced test mode
2—full-scale tests

We believe that during the tests in forced mode, there is an acceleration of thc process of changing the protective properties, that is,

$$\varepsilon_f(t) = -\alpha A \exp(-\alpha t) + \beta t \tag{6.57}$$

Then, according to Eq. (6.57), the recalculation function $x\ (t)$ can be determined from the expression

$$x(t) = -\frac{1}{\alpha}\ln\left\{\exp(-\alpha t) + \frac{\beta t^2}{2A}\right\} \tag{6.58}$$

In accordance with the theory of reliability, the probability of failure-free operation of $P\ (t)$ can be described by the exponential dependence

$$P(t) = e^{-\lambda t} \tag{6.59}$$

However, earlier studies [23–25] indicate that the aging model of coatings should take into account the components that characterize the hereditary factor. Thus it seems to us that the functions expressing the reliability indicators should also reflect the hereditary factor. Taking into account the foregoing, the probability of failure-free operation as well as the hereditary factor can be represented by a function of the form

$$P(t) = e^{-\lambda t - e^{\beta t} + 1} \tag{6.60}$$

Table 6.8 Probability I of failure-free operation of the lime coating

Day since the beginning of operation	The probability of failure-free operation of the lime coating in normal operation	
	Calculation on the experimental data in the mode $I_n(t)$	Calculation with forecasting according to test data in mode $I_f(t)$
100	0.896	0.895
180	0.801	0.796
460	0.681	0.696
1095	0.599	0.602
1825	0.503	0.507

and intensity of tests $I_f(t)$ at

$$\lambda_f(t) = \lambda + \gamma e^{\gamma t} \tag{6.61}$$

Performing a recalculation of the reliability function from the forced mode to the actual operating conditions using formula (6.61), we obtain

$$\lambda f(t) = \frac{A\exp(-\alpha t) - \beta t^2/2}{A\exp(-\alpha t) + \beta t/2}(\lambda + \gamma \exp\gamma) \tag{6.62}$$

Table 6.8 shows the calculated data on the probability of failure-free operation obtained from formula (6.62). The data obtained are in good agreement with the experimental data.

References

[1] A.D. Zimon, Adhesion of Films and Coatings, Chemistry, Moscow, 1977. 351 pp.

[2] G.M. Bartenev, Yu.S. Zuev, Strength and Destruction of Highly Elastic Materials, Chemistry, Moscow, 1964. 387 pp.

[3] G.M. Bartenev, Strength and Mechanism of Polymer Destruction, Chemistry, Moscow, 1984.

[4] M.I. Karyakina, Testing of Paint and Varnish Materials and Coatings, Chemistry, Moscow, 1988. 272 p.

[5] GOST R 9.032—74 Unified System of Protection Against Corrosion and Aging. Coatings Paint and Varnish. Groups, Specifications and Notation, Publishing House of Standards, Moscow.

[6] V.I. Loganina, A.A. Fedoseev, The Law of Probability Distribution of Trouble-Free Service of Protective and Decorative Coatings. Paint and Varnish Materials and Their Application, 2002, N10, pp. 12–14.

[7] S. Sakata, A Practical Guide to Quality Management/Translated from the 4th Japanese Edition of SI Myshkin. Ed. VI Gostyaeva, Mechanical Engineering, Moscow, 1980, 215 pp.
[8] Statistical Methods for Quality Improvement, Trans. from English. Ed. H. Kume, Finance and Statistics, Moscow, 1990, 304 pp.
[9] The GOST R 50779.30—95 Statistical Methods. Acceptance Quality Control. General Requirements, Publishing House of Standards, Moscow, 1995.
[10] GOST R 50779.50—95 Statistical Methods. Acceptance Quality Control by Variables. General Requirements, Publishing House of Standards, Moscow, 1995.
[11] E. Shindovsky, O. Shurts, Statistical Methods of Quality Management, The World, Moscow, 1976.
[12] T.A. Dubrova, The Statistical Forecasting Methods, Publishing House of Unity, Moscow, 2003. 205 pp.
[13] J.K. Belyaev, E.V. Chepurin, Fundamentals of Mathematical Statistics, Science, Moscow, 1983, p. 149.
[14] P.P. Bocharov, A.V. Pechinkin, Probability. Mathematical Statistics, Gardarica, Moscow, 1988, p. 328.
[15] V. Loganina, A.A. Fedoseev, L.P. Orentlichher, Application of Statistical Methods of Quality Management of Building Materials: Monograph, Publishing Association of Construction Universities, Moscow, 2004. 104 pp.
[16] L.P. Orentlicher, V.I. Loganina, A.A. Fedoseev, Organization of statistical acceptance control of the quality of the painted surface of building products and structures, Industrial and Civil Construction, 2004, N4, pp. 37–38.
[17] Y.-y Zhang, L.-x Li, T.-y Chen, et al., Optimization of Taguchi's on-line quality feedback control system, Proc. Inst. Mech. Eng. Part B-J. Eng. Manuf. 231 (12) (2017) 2173–2183.
[18] V.I. Loganina, L.V. Makarova, Technique of the assessment of crack resistance of the protective decorative coatings, Contemp. Eng. Sci. 7 (36) (2014) 1967–1973.
[19] V.I. Loganina, L.V. Makarova, To a technique of an assessment of crack resistance of protective and decorative coatings, Plasts No.4 (2003) 43–44.
[20] V.I. Loganina, L.V. Makarova, Evaluation of the influence of substrate quality on the crack resistance of protective and decorative coatings, Ind. Coloring N1 (2005) 53–56.
[21] N.M. Sedyakin, On a physical principle of reliability, in: Proceedings of the Academy of Sciences SSSR. Tehnicheskaya kibernetika, 1966, N3.
[22] V.A Smagie, On a model of forced testing, Reliability and Quality Control, 1966, N4.
[23] G.M. Bartenev, Strength and Fracture Mechanism of Polymers, Chemistry State Press, Moscow, 1984. 280 pp.
[24] G.M. Bartenev, Yu.S. Zuev, Strength and Destruction of Highly Elastic Materials, Chemistry, Moscow-Leningrad, 1984.
[25] V.I. Loganina, Model of aging coatings based on hereditary factors, Contemp. Eng. Sci. 8 (4) (2015). 165—170HIKARI Ltd, www.m-hikari.com https://doi.org/10.12988/ces.2015.518.

Index

Note: Page number followed by "*f*" and "*t*" refer to figures and tables, respectively.

T

CPI Antony Rowe
Eastbourne, UK
March 14, 2019